50% OFF Online HSPT Prep Course!

Dear Customer,

We consider it an honor and a privilege that you chose our HSPT Study Guide. As a way of showing our appreciation and to help us better serve you, we have partnered with Mometrix Test Preparation to offer you **50% off their online HSPT Course.** Many HSPT courses are needlessly expensive and don't deliver enough value. With their course, you get access to the best HSPT prep material, and you only pay half price.

Mometrix has structured their online course to perfectly complement your printed study guide. The HSPT Online Course contains **in-depth lessons** that cover all the most important topics, **120+ video reviews** that explain difficult concepts, over **1,500 practice questions** to ensure you feel prepared, and more than **350 digital flashcards**, so you can study while you're on the go.

Online HSPT Prep Course

Topics Covered:

- Verbal Skills
    - Synonyms and Antonyms
    - Logical and Verbal Classifications
- Reading Comprehension
    - Informational and Literary Texts
    - Figurative Language
- Language
    - Foundations of Grammar
    - Structure and Purpose of a Sentence
- Mathematics
    - Fractions, Decimals, and Percentages
    - Polygons and Triangles
- Quantitative Skills

Course Features:

- HSPT Study Guide
    - Get content that complements our best-selling study guide.
- Full-Length Practice Tests
    - With over 1,500 practice questions, you can test yourself again and again.
- Mobile Friendly
    - If you need to study on the go, the course is easily accessible from your mobile device.
- HSPT Flashcards
    - Their course includes a flashcards mode with over 350 content cards for you to study.

To receive this discount, visit their website: mometrix.com/university/hspt and add the course to your cart. At the checkout page, enter the discount code: **APEXHSPT50**

If you have any questions or concerns, please contact them at universityhelp@mometrix.com.

Sincerely,

 in partnership with

FREE

Free Study Tips DVD

In addition to the tips and content in this guide, we have created a FREE DVD with helpful study tips to further assist your exam preparation. **This FREE Study Tips DVD provides you with top-notch tips to conquer your exam and reach your goals.**

Our simple request in exchange for the strategy-packed DVD is that you email us your feedback about our study guide. We would love to hear what you thought about the guide, and we welcome any and all feedback—positive, negative, or neutral. It is our #1 goal to provide you with top quality products and customer service.

To receive your **FREE Study Tips DVD**, email freedvd@apexprep.com. Please put "FREE DVD" in the subject line and put the following in the email:

> a. The name of the study guide you purchased.
>
> b. Your rating of the study guide on a scale of 1-5, with 5 being the highest score.
>
> c. Any thoughts or feedback about your study guide.
>
> d. Your first and last name and your mailing address, so we know where to send your free DVD!

Thank you!

HSPT Prep Book for Catholic High School Entrance Exams

HSPT Practice Questions and Study Guide
[2nd Edition]

Matthew Lanni

Table of Contents

Test Taking Strategies

1. Reading the Whole Question

A popular assumption in Western culture is the idea that we don't have enough time for anything. We speed while driving to work, we want to read an assignment for class as quickly as possible, or we want the line in the supermarket to dwindle faster. However, speeding through such events robs us from being able to thoroughly appreciate and understand what's happening around us. While taking a timed test, the feeling one might have while reading a question is to find the correct answer as quickly as possible. Although pace is important, don't let it deter you from reading the whole question. Test writers know how to subtly change a test question toward the end in various ways, such as adding a negative or changing focus. If the question has a passage, carefully read the whole passage as well before moving on to the questions. This will help you process the information in the passage rather than worrying about the questions you've just read and where to find them. A thorough understanding of the passage or question is an important way for test takers to be able to succeed on an exam.

2. Examining Every Answer Choice

Let's say we're at the market buying apples. The first apple we see on top of the heap may *look* like the best apple, but if we turn it over we can see bruising on the skin. We must examine several apples before deciding which apple is the best. Finding the correct answer choice is like finding the best apple. Although it's tempting to choose an answer that seems correct at first without reading the others, it's important to read each answer choice thoroughly before making a final decision on the answer. The aim of a test writer might be to get as close as possible to the correct answer, so watch out for subtle words that may indicate an answer is incorrect. Once the correct answer choice is selected, read the question again and the answer in response to make sure all your bases are covered.

3. Eliminating Wrong Answer Choices

Sometimes we become paralyzed when we are confronted with too many choices. Which frozen yogurt flavor is the tastiest? Which pair of shoes look the best with this outfit? What type of car will fill my needs as a consumer? If you are unsure of which answer would be the best to choose, it may help to use process of elimination. We use "filtering" all the time on sites such as eBay® or Craigslist® to eliminate the ads that are not right for us. We can do the same thing on an exam. Process of elimination is crossing out the answer choices we know for sure are wrong and leaving the ones that might be correct. It may help to cover up the incorrect answer choice. Covering incorrect choices is a psychological act that alleviates stress due to the brain being exposed to a smaller amount of information. Choosing between two answer choices is much easier than choosing between all of them, and you have a better chance of selecting the correct answer if you have less to focus on.

4. Sticking to the World of the Question

When we are attempting to answer questions, our minds will often wander away from the question and what it is asking. We begin to see answer choices that are true in the real world instead of true in the world of the question. It may be helpful to think of each test question as its own little world. This world may be different from ours. This world may know as a truth that the chicken came before the egg or may assert that two plus two equals five. Remember that, no matter what hypothetical nonsense may be in the question, assume it to be true. If the question states that the chicken came before the egg, then choose your answer based on that truth. Sticking to the world of the question means placing all of our biases and assumptions aside and relying on the question to guide us to the correct answer. If we are simply looking for answers that are correct based on our own judgment, then we may choose incorrectly. Remember an answer that is true does not necessarily answer the question.

5. Key Words

If you come across a complex test question that you have to read over and over again, try pulling out some key words from the question in order to understand what exactly it is asking. Key words may be words that surround the question, such as *main idea, analogous, parallel, resembles, structured,* or *defines*. The question may be asking for the main idea, or it may be asking you to define something. Deconstructing the sentence may also be helpful in making the question simpler before trying to answer it. This means taking the sentence apart and obtaining meaning in pieces, or separating the question from the foundation of the question. For example, let's look at this question:

> Given the author's description of the content of paleontology in the first paragraph, which of the following is most parallel to what it taught?

The question asks which one of the answers most *parallels* the following information: The *description* of paleontology in the first paragraph. The first step would be to see *how* paleontology is described in the first paragraph. Then, we would find an answer choice that parallels that description. The question seems complex at first, but after we deconstruct it, the answer becomes much more attainable.

6. Subtle Negatives

Negative words in question stems will be words such as *not, but, neither,* or *except*. Test writers often use these words in order to trick unsuspecting test takers into selecting the wrong answer—or, at least, to test their reading comprehension of the question. Many exams will feature the negative words in all caps (*which of the following is NOT an example*), but some questions will add the negative word seamlessly into the sentence. The following is an example of a subtle negative used in a question stem:

> According to the passage, which of the following is *not* considered to be an example of paleontology?

If we rush through the exam, we might skip that tiny word, *not,* inside the question, and choose an answer that is opposite of the correct choice. Again, it's important to read the question fully, and double check for any words that may negate the statement in any way.

7. Spotting the Hedges

The word "hedging" refers to language that remains vague or avoids absolute terminology. Absolute terminology consists of words like *always, never, all, every, just, only, none,* and *must*. Hedging refers to words like *seem, tend, might, most, some, sometimes, perhaps, possibly, probability,* and *often*. In some cases, we want to choose answer choices that use hedging and avoid answer choices that use absolute terminology. It's important to pay attention to what subject you are on and adjust your response accordingly.

8. Restating to Understand

Every now and then we come across questions that we don't understand. The language may be too complex, or the question is structured in a way that is meant to confuse the test taker. When you come across a question like this, it may be worth your time to rewrite or restate the question in your own words in order to understand it better. For example, let's look at the following complicated question:

> Which of the following words, if substituted for the word *parochial* in the first paragraph, would LEAST change the meaning of the sentence?

Let's restate the question in order to understand it better. We know that they want the word *parochial* replaced. We also know that this new word would "least" or "not" change the meaning of the sentence. Now let's try the sentence again:

> Which word could we replace with *parochial,* and it would not change the meaning?

Restating it this way, we see that the question is asking for a synonym. Now, let's restate the question so we can answer it better:

> Which word is a synonym for the word *parochial*?

Before we even look at the answer choices, we have a simpler, restated version of a complicated question.

9. Predicting the Answer

After you read the question, try predicting the answer *before* reading the answer choices. By formulating an answer in your mind, you will be less likely to be distracted by any wrong answer choices. Using predictions will also help you feel more confident in the answer choice you select. Once you've chosen your answer, go back and reread the question and answer choices to make sure you have the best fit. If you have no idea what the answer may be for a particular question, forego using this strategy.

10. Avoiding Patterns

One popular myth in grade school relating to standardized testing is that test writers will often put multiple-choice answers in patterns. A runoff example of this kind of thinking is that the most common answer choice is "C," with "B" following close behind. Or, some will advocate certain made-up word patterns that simply do not exist. Test writers do not arrange their correct answer choices in any kind of pattern; their choices are randomized. There may even be times where the correct answer choice will be the same letter for two or three questions in a row, but we have no way of knowing when or if this might happen. Instead of trying to figure out what choice the test writer probably set as being correct, focus on what the *best answer choice* would be out of the answers you are presented with. Use the tips above, general knowledge, and reading comprehension skills in order to best answer the question, rather than looking for patterns that do not exist.

FREE DVD OFFER

Achieving a high score on your exam depends not only on understanding the content, but also on understanding how to apply your knowledge and your command of test taking strategies. **Because your success is our primary goal, we offer a FREE Study Tips DVD, which provides top-notch test taking strategies to help you optimize your testing experience.**

Our simple request in exchange for the strategy-packed DVD is that you email us your feedback about our study guide.

To receive your **FREE Study Tips DVD**, email freedvd@apexprep.com. Please put "FREE DVD" in the subject line and put the following in the email:

a. The name of the study guide you purchased.

b. Your rating of the study guide on a scale of 1-5, with 5 being the highest score.

c. Any thoughts or feedback about your study guide.

d. Your first and last name and your mailing address, so we know where to send your free DVD!

Introduction

Function of the Test

The High School Placement Test (HSPT) is a Catholic high school entrance exam for eighth grade students going into ninth grade that assists with scholarship selection, curriculum placement, and admission into Catholic high schools. The HSPT is a nationwide test that has been administered for over fifty years. Each year, over 100,000 students take the HSPT.

Test Administration

Individual schools or dioceses order and administer the HSPT, so make sure to contact your student's school to learn the details of which placement test is administered by their school and what their registration process is like. Usually, students are only allowed to take the HSPT once. Some exceptions have been made in the past, but it is ultimately up to the school whether or not your student may retest.

Regarding special accommodations for your student, it is recommended that parents contact their student's school, as the school will offer special accommodations based on their administration's own individual policies.

Test Format

Although the HSPT is given nationwide, the testing environment will vary, as your student will register and take the exam at the school they attend. The HSPT is structured as a five-part exam with 298 questions. Students have two hours and thirty minutes to complete the test. The table below describes the test content along with the number of questions.

Content	Questions
Verbal	60 questions
Quantitative	52 questions
Reading	62 questions
Mathematics	64 questions
Language	60 questions

The verbal section of the HSPT includes analogies, logic, classifications, synonyms, and antonyms. The quantitative section of the HSPT includes number series and manipulation, and geometric and non-geometric comparisons. The reading section of the HSPT includes comprehension and vocabulary. The mathematics section includes math concepts and problem-solving. The language section of the HSPT includes composition, spelling, punctuation, capitalization, and usage.

Scoring

The HSPT is scored by giving students one point for every answer they guess correctly. Points are not deducted for incorrect answers. Making an educated guess is encouraged; always answer the question, even if you are not sure the answer is correct. The score report will include the raw score, scaled score (200–800), the percentile rank, the cognitive skills (total score from verbal and quantitative skills) and basic skills (total score from reading, mathematics, and language skills).

Verbal Skills

Verbal Analogies

The analogy section is designed to test your knowledge of word definitions as well as their relationship to one another. Making analogies is part of an important cognitive process that acts as the basis of metaphor and association. Analogical thinking is used in problem solving, creative thinking, argumentation, invention, communication, and memory, among other intellectual operations. **Verbal analogies** are a way to determine associations between objects, their signifying words, and each word's specific connotations. Learning verbal analogies is valued among many standardized exams because of its usefulness in language and learning, especially among figurative language such as metaphors, similes, and allegories.

Question Format

Below we will look at the different types of analogies, but it's also helpful to know what format the analogy section uses and how best to answer the questions. The question will give you either two words or three words to work with. Let's start with an example with only two words:

Cat is to **mammal** as

a. **shoe** is to **foot.**
b. **kitten** is to **feline**.
c. **dolphin** is to **amphibian**.
d. **lizard** is to **reptile.**
e. **lamp** is to **bedroom**.

This type of question requires you to carefully study the relationship between the first pair of words. The first pair of words have the relationship of **category and type.** We see that in the world of mammals, a cat is a **type** of mammal, or can be considered a mammal. The word *cat* falls within the classification of *mammal.* Now when we look at the answer choices, we have to determine that the first word falls within an appropriate category of the second word. *Shoe* is not a category of *foot,* so Choice *A* is incorrect. *Feline* is a synonym of *cat,* or is part of the cat family, so this is not the best answer. Choice *B*'s *kitten* and *feline* are too close to each other in meaning. Choice *C* is incorrect because a *dolphin* does not fall under the class of *amphibian*. Rather, it is a mammal. Choice *D* looks like the correct answer. *Lizard* is a type of *reptile* and falls under the category of reptile. This pair of words has the same relationship as the original pair. Finally, Choice *E* is incorrect. A *lamp* may reside in the *bedroom,* but the association is too generalized for this question.

Now let's look at a question with three words to start off with:

Heat is to **scorching** as **cold** is to

a. burning.
b. freezing.
c. melting.
d. ice.
e. Alaska.

In the questions where we are given three terms, we still need to identify the relationship between the first pair of words. Here, we see that this type of analogy is a **degree of intensity**. We are given the word "hot," and we know that "scorching" means "really, really hot." Let's look at what cold's intensity looks like. Choices *A* and *C*, *burning* and *melting*, are the opposite of an intensity of cold, so these are incorrect. Choice *B* looks like a good answer. *Freezing* is an intensity of cold, so let's mark this as correct. Choice *D*, *ice*, is an object that is really cold, and Choice *E*, *Alaska*, is a place that is really cold. These could be in the running. However, to narrow down between *freezing*, *ice*, and *Alaska*, it's important to look at the closest analogy to the word *scorching*. The word is an adjective, not a noun or proper noun, as *ice* and *Alaska* are. Since *scorching* and *freezing* both occupy the same part of speech, let's choose *freezing*, Choice *B*, as the correct answer.

Types of Analogies

In its most basic form, an analogy compares two different things. An analogy is a pair of words that parallels the situation or relationship given in another pair of words. The table below gives examples of the different types of analogies you may see:

Type of Analogy	Relationship	Example
Synonym	The pair of words are alike in meaning.	**Happy** is to **joyous** as **sad** is to **somber**.
Antonym	The pair of words are opposite in meaning.	**Lucky** is to **unfortunate** as **victorious** is to **defeated**.
Part to Whole	One word stands for a whole, and the other word stands as a part to that whole.	**Chapter** is to **novel** as **pupil** is to **eye**.
Category/Type	One thing belongs in a category of another thing.	**Screwdriver** is to **tool** as **apartment** is to **dwelling**.
Object to Function	A pair depicting a tool and the use of that tool.	**Shovel** is to **dig** as **oven** is to **bake**.
Degree of Intensity	The pair of words shows a difference in degree.	**Funny** is to **hysterical** as **interest** is to **adoration**.
Cause and Effect	The pair shows that one word is created by the other word.	**Hard work** is to **success** as **privilege** is to **comfort**.
Symbol and Representation	A word and its representation in the context of a culture.	**Rose** is to **love** as **flag** is to **patriotism**.
Performer to Related Action	A person and their related action.	**Professor** is to **teach** as **doctor** is to **heal**.

Synonyms and Antonyms

The **Synonym questions** on the HSPT are designed to assess the test taker's vocabulary knowledge and skill in determining the answer choice with a meaning that most nearly matches that of the word presented in the question. In this way, these questions task test takers with applying their vocabulary skills, understanding language, and utilizing logic to select the best synonyms. The **Antonym questions** on the HSPT are also designed to assess the test taker's vocabulary knowledge by determining which answer choice means the opposite of the given word.

Question Format

The questions in this section are constructed very simply: the prompt is a single word, which is then followed by the five answer choices, each of which is also a single word. Test takers must consider the definition or meaning of the word provided in the prompt, and then select the answer choice that most nearly means the same thing. Consider the following example of a question:

SERENE
    a. Serious
    b. Calm
    c. Tired
    d. Nervous
    e. Jealous

Test takers must consider the meaning of the given word (*serene*), read the five potential definitions or possible synonyms, and pick the word that most closely means the same thing as the word in the prompt. In this case, Choice *B* is the best answer because *serene* means calm.

Analyzing Word Parts

By learning some of the **etymologies** of words and their parts, readers can break new words down into components and analyze their combined meanings. For example, the root word *soph* is Greek for wise or knowledge. Knowing this informs the meanings of English words including *sophomore, sophisticated,* and *philosophy.* Those who also know that *phil* is Greek for love will realize that *philosophy* means the love of knowledge. They can then extend this knowledge of *phil* to understand *philanthropist* (one who loves people), *bibliophile* (book lover), *philharmonic* (loving harmony), *hydrophilic* (water-loving), and so on. In addition, *phob-* derives from the Greek *phobos,* meaning fear. This informs all words ending with it as meaning fear of various things: *acrophobia* (fear of heights), *arachnophobia* (fear of spiders), *claustrophobia* (fear of enclosed spaces), *ergophobia* (fear of work), and *hydrophobia* (fear of water), among others.

Some words that originate from other languages, like ancient Greek, are found in large numbers and varieties of English words. An advantage of the shared ancestry of these words is that once readers recognize the meanings of some Greek words or word roots, they can determine or at least get an idea of what many different English words mean. As an example, the Greek word *métron* means to measure, a measure, or something used to measure; the English word *meter* derives from it. Knowing this informs many other English words, including *altimeter, barometer, diameter, hexameter, isometric,* and *metric.* While readers must know the meanings of the other parts of these words to decipher their meaning fully, they already have an idea that they are all related in some way to measures or measuring.

While all English words ultimately derive from a proto-language known as Indo-European, many of them historically came into the developing English vocabulary later, from sources like the ancient Greeks' language, the Latin used throughout Europe and much of the Middle East during the reign of the Roman Empire, and the Anglo-Saxon languages used by England's early tribes. In addition to classic revivals and native foundations, by the Renaissance era, other influences included French, German, Italian, and Spanish. Today we can often discern English word meanings by knowing common roots and affixes, particularly from Greek and Latin.

The following is a list of common **prefixes** and their meanings:

Prefix	Definition	Examples
a-	without	atheist, agnostic
ad-	to, toward	advance
ante-	before	antecedent, antedate
anti-	opposing	antipathy, antidote
auto-	self	autonomy, autobiography
bene-	well, good	benefit, benefactor
bi-	two	bisect, biennial
bio-	life	biology, biosphere
chron-	time	chronometer, synchronize
circum-	around	circumspect, circumference
com-	with, together	commotion, complicate
contra-	against, opposing	contradict, contravene
cred-	belief, trust	credible, credit
de-	from	depart
dem-	people	demographics, democracy
dis-	away, off, down, not	dissent, disappear
equi-	equal, equally	equivalent
ex-	former, out of	extract
for-	away, off, from	forget, forswear
fore-	before, previous	foretell, forefathers
homo-	same, equal	homogenized
hyper-	excessive, over	hypercritical, hypertension
in-	in, into	intrude, invade
inter-	among, between	intercede, interrupt
mal-	bad, poorly, not	malfunction
micr-	small	microbe, microscope
mis-	bad, poorly, not	misspell, misfire
mono-	one, single	monogamy, monologue
mor-	die, death	mortality, mortuary
neo-	new	neolithic, neoconservative
non-	not	nonentity, nonsense
omni-	all, everywhere	omniscient
over-	above	overbearing
pan-	all, entire	panorama, pandemonium
para-	beside, beyond	parallel, paradox
phil-	love, affection	philosophy, philanthropic
poly-	many	polymorphous, polygamous
pre-	before, previous	prevent, preclude
prim-	first, early	primitive, primary
pro-	forward, in place of	propel, pronoun
re-	back, backward, again	revoke, recur
sub-	under, beneath	subjugate, substitute

super-	above, extra	supersede, supernumerary
trans-	across, beyond, over	transact, transport
ultra-	beyond, excessively	ultramodern, ultrasonic, ultraviolet
un-	not, reverse of	unhappy, unlock
vis-	to see	visage, visible

The following is a list of common **suffixes** and their meanings:

Suffix	Definition	Examples
-able	likely, able to	capable, tolerable
-ance	act, condition	acceptance, vigilance
-ard	one that does excessively	drunkard, wizard
-ation	action, state	occupation, starvation
-cy	state, condition	accuracy, captaincy
-er	one who does	teacher
-esce	become, grow, continue	convalesce, acquiesce
-esque	in the style of, like	picturesque, grotesque
-ess	feminine	waitress, lioness
-ful	full of, marked by	thankful, zestful
-ible	able, fit	edible, possible, divisible
-ion	action, result, state	union, fusion
-ish	suggesting, like	churlish, childish
-ism	act, manner, doctrine	barbarism, socialism
-ist	doer, believer	monopolist, socialist
-ition	action, result, state,	sedition, expedition
-ity	quality, condition	acidity, civility
-ize	cause to be, treat with	sterilize, mechanize, criticize
-less	lacking, without	hopeless, countless
-like	like, similar	childlike, dreamlike
-ly	like, of the nature of	friendly, positively
-ment	means, result, action	refreshment, disappointment
-ness	quality, state	greatness, tallness
-or	doer, office, action	juror, elevator, honor
-ous	marked by, given to	religious, riotous
-some	apt to, showing	tiresome, lonesome
-th	act, state, quality	warmth, width
-ty	quality, state	enmity, activity

The following is a list of **root words** and their meanings:

Root	Definition	Examples
ambi	both	ambidextrous, ambiguous
anthropo	man; humanity	anthropomorphism, anthropology
auto	self	automobile, autonomous
bene	good	benevolent, benefactor
bio	life	biology, biography
chron	time	chronology
circum	around	circumvent, circumference
dyna	power	dynasty, dynamite
fort	strength	fortuitous, fortress
graph	writing	graphic
hetero	different	heterogeneous
homo	same	homonym, homogenous
hypo	below, beneath	hypothermia
morph	shape; form	morphology
mort	death	mortal, mortician
multi	many	multimedia, multiplication
nym	name	antonym, synonym
phobia	fear	claustrophobia
port	carry	transport
pseudo	false	pseudoscience, pseudonym
scope	viewing instrument	telescope, microscope
techno	art; science; skill	technology, techno
therm	heat	thermometer, thermal
trans	across	transatlantic, transmit
under	too little	underestimate

Logic

The ten **Logic questions** on the HSPT exam present a hypothetical scenario—typically in the form of a syllogism—and then the same three answer choices: "true," "false," and "uncertain." Test takers must use their deductive reasoning skills to evaluate the scenario and choose the correct answer. The logic arguments on the test should be taken to be true, even if they may not hold up in the real world.

It is often helpful to diagram the logic simulation to help determine the conclusion. Let's consider an example:

Chocolate chip cookies are more popular than sugar cookies. Sugar cookies are less popular than snickerdoodles. Oatmeal cookies are more popular than chocolate chip cookies. If the first two statements are true, the third statement is:

a. true
b. false
c. uncertain

The correct answer is *C*, uncertain, because we don't have enough information to definitively determine whether the third statement is true or false. It is helpful to diagram the scenario on a piece of paper. Draw a circle with a "C" in it to represent chocolate chip cookies. Draw another circle, lower down on the paper, with an "S" in the middle for sugar cookies, since the sugar cookies are not as popular as the chocolate chip cookies. We are told this statement is true. For the second statement, we are told that sugar cookies, our "S" circle, are less popular than oatmeal cookies, which we can represent with an "O" circle. This means that the "S" must be somewhere below the "O," but is the "O" above or below the "C"? Because we are only told the first two statements are true, we don't have enough information to determine whether the third statement—that oatmeal cookies are more popular than chocolate chip ones—is true or not. At this point, it's uncertain.

Let's try another example of a logic question:

All seals have flippers. Some seals live in the Arctic. Some Arctic animals have flippers.

a. true
b. false
c. uncertain

This time, we can determine that the answer is Choice *A, true*. If all seals have flippers, and some of those seals live in the Arctic, then it is true that some Arctic animals (at least the seals) have flippers.

What if, however, the third sentence said: "all Arctic animals have flippers"? Then our answer would have to be Choice *C, uncertain*. Why would it be "uncertain" and not "false"? The reason is that in the world of this simulation, we don't *know* whether all the Arctic animals have flippers or not. If we had "All seals have flippers. Some seals live in the Arctic. All Arctic animals have flippers," then we don't know what other Arctic animals besides seals are included in the discussion. Even though we can name other Arctic animals in the real world that don't have scales (for example, polar bears), it's important to remember that the logic questions must be answered not based on outside knowledge, but rather only on the statements given in the question. Therefore, in the world of our hypothetical stimulus that ends in "All Arctic animals have flippers," we aren't told what other animals are included and what their feet are like. Therefore, the answer would have to be "uncertain."

The important takeaways are to read each question carefully and to consider only what is said in the world of the question rather than what you might know about the real world. Two of the most helpful strategies are to diagram the problem to create a visual representation and to assign possible values (such as heights or ages), where possible, to get a more tangible grasp on the provided information.

Verbal Classification

For the **Verbal Classification questions**, test takers are presented with the question: "Which word does not belong with the others?" This standard question is followed by four answer choices, and the test taker must determine the "odd one out," or the one choice that is different from the others in some way. While the instructions are simple, there is no way to guarantee a perfect score on this section simply because there is no official master list of potential words that might appear on the test. As such, even with a vast vocabulary, it's possible that test takers may encounter unfamiliar terms in the Verbal Classification section. The test takers who excel on this section have prepared by honing the same type of word knowledge skills required in the Synonyms and Antonyms sections, in addition to building their vocabulary directly through explicit studying and indirectly through reading texts of all types.

On the test, if an unfamiliar word is encountered, test takers can try to determine the root and affixes to help in deriving the word's meaning. Additionally, it's useful to discern if the word has any negative or positive connotations, and whether this differs or parallels the general connotation of the other words in the set. It's important to remember that test takers are not tasked with selecting the actual definitions of the word; they only need to identify the one word that is unlike the other three. Oftentimes, just having a feel for the general meaning or "mood" of the words can be enough to answer the question correctly.

Let's look at an example:

Which word does not belong with the others?
> a. Afraid
> b. Scared
> c. Occupied
> d. Frightened

In this example, we can determine *occupied* to be the word that doesn't belong. *Afraid, scared,* and *frightened* are near synonyms. *Occupied* is unrelated.

Some verbal classification questions contain three synonyms and then one word that doesn't have the same or similar meaning to the others like the example just shown. Other verbal classification questions provide classifications of objects or people, such as workers/professions or types of flowers. In these categorical types of questions, test takers must select the one object or type of person that doesn't belong. For example, one with types of flowers may have three types of flowers and one tree species. Therefore, the tree doesn't belong. Let's look at the following question involving professions:

Which word does not belong with the others?
> a. Rheumatologist
> b. Dentist
> c. Cardiologist
> d. Pediatrician

Remember that it is not imperative that you know all of the definitions of the words in the set in order to correctly answer this question. As with the questions on the real test, we can easily use the process of elimination to answer this example question. Let's assume everyone knows what a dentist is. Then let's say you know that a pediatrician is a medical doctor who specializes in the care of children because you remember going to one when you were younger, or perhaps you still see one. Then consider the word *cardiologist.* Test takers who have prepared by studying roots and affixes will be able to recognize the common prefix *cardio-*, which relates to the heart. Similarly, the suffix *-ologist* is used to mean someone who studies a particular subject. Thus, the term *cardiologist* is used to identify heart doctors. Let's say you don't know the meaning of the word *rheumatologist*. It's still possible to answer the question correctly by employing logic and word knowledge skills. The term *rheumatologist* is similar to many other types of doctors—not only cardiologist, which is in this set—but others like ophthalmologist, endocrinologist, and gastroenterologist. *Rheum-* is a prefix in words like *rheumatoid arthritis* and *rheumatic fever*, so even if you don't know that the prefix means joints, and a rheumatologist is a joint doctor, if you've heard of these or other rheum-based medical words, you can assume that a rheumatologist is some sort of doctor. Therefore, the choices become three types of medical doctors and a dentist. Thus, the "odd one out" is Choice *B, dentist.* A dentist is not a medical doctor.

Practice Questions

Verbal Analogies

The questions below ask you to find relationships between words. For each question, select the answer that best completes the meaning of the sentence.

1. Chapter is to book as
    a. Book is to story.
    b. Fable is to myth.
    c. Paragraph is to essay.
    d. Dialogue is to play.

2. Dress is to garment as
    a. Diesel is to fuel.
    b. Suit is to tie.
    c. Clothing is to wardrobe.
    d. Coat is to winter.

3. Car is to garage as plane is to
    a. Sky.
    b. Hangar.
    c. Airport.
    d. Runway.

4. Trickle is to gush as
    a. Bleed is to cut.
    b. Rain is to snow.
    c. Tepid is to scorching.
    d. Sob is to sniffle.

5. Obviate is to preclude as
    a. Exclude is to include.
    b. Conceal is to avert.
    c. Pontificate is to ponder.
    d. Appease is to placate.

6. Acre is to area as fathom is to
    a. Depth.
    b. Angle degree.
    c. Wind speed.
    d. Width.

7. Nadir is to zenith as valley is to
    a. Depression.
    b. Pinnacle.
    c. Climb.
    d. Rise.

8. Blueprint is to architect as
    a. Stethoscope is to doctor.
    b. Score is to composer.
    c. Lathe is to craftsman.
    d. Outline is to drawing.

9. Adore is to appreciate as loathe is to
    a. Detest.
    b. Hate.
    c. Appreciate.
    d. Dislike.

10. Painter is to easel as weaver is to
    a. Tapestry.
    b. Yarn.
    c. Needle.
    d. Loom.

Synonyms

Each of the questions below has one word. The one word is followed by four words or phrases. Please select one answer whose meaning is closest to the word in capital letters.

11. REPROACH
    a. Locate
    b. Blame
    c. Concede
    d. Orate

12. MILIEU
    a. Bacterial
    b. Damp
    c. Ancient
    d. Environment

13. GUILE
    a. Deception
    b. Stubborn
    c. Naïve
    d. Gullible

14. ASSENT
    a. Acquiesce
    b. Climb
    c. Assert
    d. Demand

15. DEARTH
    a. Grounded
    b. Scarcity
    c. Lethal
    d. Risky

16. CONSPICUOUS
    a. Scheme
    b. Obvious
    c. Secretive
    d. Ballistic

17. ONEROUS
    a. Responsible
    b. Generous
    c. Hateful
    d. Burdensome

18. BANAL
    a. Inane
    b. Novel
    c. Painful
    d. Complimentary

19. CONTRITE
    a. Tidy
    b. Unrealistic
    c. Remorseful
    d. Corrupt

20. MOLLIFY
    a. Pacify
    b. Blend
    c. Negate
    d. Amass

21. CAPRICIOUS
    a. Skillful
    b. Sanguine
    c. Chaotic
    d. Fickle

22. PALTRY
    a. Appealing
    b. Worthy
    c. Trivial
    d. Fancy

23. SHIRK
   a. Counsel
   b. Evade
   c. Diminish
   d. Sharp

24. ASSUAGE
   a. Irritate
   b. Persuade
   c. Soothe
   d. Redirect

25. TACIT
   a. Unspoken
   b. Shortened
   c. Tenuous
   d. Regal

Logic

26. At a certain bakery, a croissant is cheaper than a bagel. Oatmeal is more expensive than a bagel. Oatmeal costs less than a croissant. If the first two statements are true, the third is:
   a. true
   b. false
   c. uncertain

27. Amelia has more siblings than Jude. Jude has more siblings than Kendra. Kendra has more siblings than Amelia. If the first two statements are true, the third is:
   a. true
   b. false
   c. uncertain

28. Mysteries are better than romances. Fantasies are not as good as mysteries. Fantasies are better than romances. If the first two statements are true, the third is:
   a. true
   b. false
   c. uncertain

29. All soups at Mimi's diner contain chicken stock. All chicken stock is considered to be an animal product and not part of a vegetarian diet. Only some soups at Mimi's contain animal products. If the first two statements are true, the third is:
   a. true
   b. false
   c. uncertain

30. Ben is not overweight. Ben is a runner. No runners are overweight. If the first two statements are true, the third is:
   a. true
   b. false
   c. uncertain

31. Lollipops are sweeter than jolly ranchers. Candy buttons are sweeter than lollipops. Jolly ranchers are not as sweet as candy buttons. If the first two statements are true, the third is:
  a. true
  b. false
  c. uncertain

32. Some socks in Jody's bin contain wool. All socks in the bin have cotton. No socks are wool and cotton. If the first two statements are true, the third is:
  a. true
  b. false
  c. uncertain

33. No made-for-TV movies are entertaining. Lifetime movies are made-for-TV movies. Lifetime movies are not entertaining. If the first two statements are true, the third is:
  a. true
  b. false
  c. uncertain

34. George drives further to work than Bill. Bill does not drive as far to work as Seth. Seth and George drive the same distance to work. If the first two statements are true, the third is:
  a. true
  b. false
  c. uncertain

35. Some cars are considered to be eco-friendly vehicles. All eco-friendly vehicles are eligible to receive a rebate from the EPA. Some cars are eligible to receive a rebate from the EPA. If the first two statements are true, the third is:
  a. true
  b. false
  c. uncertain

Verbal Classifications

In the following questions, one word does not belong with the others. Your answer choice will reflect the word that is the "odd one out."

36. Which word does *not* belong with the others?
  a. Bleach
  b. Vinegar
  c. Lemon Juice
  d. Urine

37. Which word does *not* belong with the others?
  a. Trout
  b. Cod
  c. Shrimp
  d. Flounder

38. Which word does *not* belong with the others?
    a. Hexagon
    b. Sphere
    c. Trapezoid
    d. Triangle

39. Which word does *not* belong with the others?
    a. Funny
    b. Nervously
    c. Intelligent
    d. Sticky

40. Which word does *not* belong with the others?
    a. Iron
    b. Stone
    c. Ice
    d. Platinum

41. Which word does *not* belong with the others?
    a. Urgent
    b. Dire
    c. Prudent
    d. Exigent

42. Which word does *not* belong with the others?
    a. Social worker
    b. Counselor
    c. Psychologist
    d. Physical therapist

43. Which word does *not* belong with the others?
    a. Schedule
    b. Defer
    c. Table
    d. Postpone

44. Which word does *not* belong with the others?
    a. Ponder
    b. Mull
    c. Utilize
    d. Deliberate

45. Which word does *not* belong with the others?
    a. Ate
    b. Discovered
    c. Cook
    d. Kicked

46. Which word does *not* belong with the others?
    a. Cornea
    b. Iris
    c. Fovea
    d. Sinus

47. Which word does *not* belong with the others?
    a. Bear
    b. Carry
    c. Glare
    d. Their

48. Which word does *not* belong with the others?
    a. Attractive
    b. Alluring
    c. Captivating
    d. Capitalizing

49. Which word does *not* belong with the others?
    a. Dim
    b. Spot
    c. Dull
    d. Faint

50. Which word does *not* belong with the others?
    a. Rifle
    b. Concinnate
    c. Scour
    d. Rummage

51. Which word does *not* belong with the others?
    a. Foal
    b. Calf
    c. Sheep
    d. Lamb

Antonyms

52. WARY
    a. Cautious
    b. Eager
    c. Tired
    d. Negligent

53. PARSIMONIOUS
   a. Lavish
   b. Miserly
   c. Harmonious
   d. Generous

54. PROPRIETY
   a. Ownership
   b. Inappropriateness
   c. Patented
   d. Abstinence

55. BOON
   a. Cacophony
   b. Hopeful
   c. Tranquility
   d. Misfortune

56. STRIFE
   a. Plague
   b. Accord
   c. Dissension
   d. Eliminate

57. QUALM
   a. Certainty
   b. Uneasiness
   c. Promote
   d. Demean

58. VALOR
   a. Commonplace
   b. Coveted
   c. Leadership
   d. Cowardice

59. ZEAL
   a. Craziness
   b. Apathy
   c. Fervor
   d. Opposition

60. EXTOL
   a. Glorify
   b. Castigate
   c. Vigilant
   d. Torch

Answer Explanations

1. C: This is a part/whole analogy. A chapter is a section, or portion, of a book. All of the choices relate to literary topics, but the best option is paragraph is to essay. A paragraph is a section, or building block, of an essay in much the same way that a chapter is in a book. The word pairs in Choices *A* and *B* are best described as near synonyms, but not necessarily parts of one another. Choice *D*, dialogue is to play, does include more of a part to the whole relationship, but dialogue is the way the story is conveyed in a play (like sentences in a book). A better matching analogy to chapter is to book would be scene or act is to play, since plays are divided into scenes or larger acts.

2. A: This is a type of category analogy. A dress is a type of garment. Garment is the broad category, and dress is the specific example used. Diesel is a type of fuel, so it holds the same relationship. A suit is not a type of tie, but it might be worn with a tie. Clothing is not a type of wardrobe; it is stored in a wardrobe. Lastly, a coat isn't a type of winter; it is a garment worn in the winter.

3. B: This is a provider/provision analogy. The analogy focuses on where the mode of transportation is stored or housed when not in use. A garage is where a car is sheltered and kept when not in use, much like a hangar for an airplane. Choice *C*, airport, might be an appealing choice, but an airport doesn't have as precise of a relationship to a plane as does a garage to a car. An airport might be where planes are located before use or where one might see a lot of planes, like a parking lot for a car.

4. C: This is an intensity analogy. Fluid that is trickling is barely moving or of low volume, while gushing fluid is moving fast and often in a larger volume. The only choices that are related by intensity are Choices *C* and *D*, but *D* reverses the relationship. Sob is a more significant cry versus a small sniffle. Tepid is lukewarm or slightly warm, while scorching is very hot.

5. D: This is a synonyms analogy, which matches terms based on their similar meanings. *Obviate* is a verb that can mean to prevent or avoid, or to render something unnecessary. It is usually used with an object, one that is often a difficulty or disadvantage. For example, wearing a helmet while cycling can obviate the risk of a skull injury should a fall occur. *Preclude* is also a verb. It means to prevent something from occurring or existing. For example, a thunderstorm may preclude a picnic. It can also mean to exclude from something. For example, an inability to get wet after surgery would preclude the patient from swimming. Of the answer choices provided, Choice *D*, appease and placate, are the only other synonyms. Like the terms in the prompt, these words are both verbs. *Placate* means to pacify, calm, or make a person less angry or upset. Similarly, *appease* means to pacify or calm someone by fulfilling their demands.

6. A: This can be considered a type of characteristic analogy because it is matching a unit of measure with an example of what that unit measures. An *acre* is an example of a unit of measure for area. Plots of land can be measured in acres, and this value gives information about how much area of land that plot occupies. A *fathom* is an example of a unit to measure depth. It is typically used to refer to water depth.

7. B: This is an antonyms analogy. *Zenith* and *nadir* are astronomical nouns with opposite meanings that have also been adopted into conversational (non-technical) English. In an astronomical sense, the *zenith* is the point in the celestial sphere or sky that lies directly above the observer, while the *nadir* is the point directly below the observer. These terms have been incorporated into common language to mean the very top or culminating point of something (the *zenith*), and the very bottom, lowest, or worst (the *nadir*). For example, the *zenith* of triathlete's athletic career might be winning the Ironman World Championships, while the *nadir* might be crashing his or her bike during a race and fracturing a bone. In this question, the term *valley* is provided for the next pair. Like *nadir,* a *valley* is a very low point. The correct answer will then be a high point, which is best captured by Choice *B,* pinnacle.

8. B: This analogy matches an occupation with what someone in that job creates and uses as a plan for their work. An architect creates a blueprint to be a rendering of the plan for the structures that builders will use to erect the building, much like a composer creates a score that musicians will follow to play the music. The other answer choices do not maintain this same relationship.

9. D: This is an intensity analogy. *Adore* is a stronger version of *appreciate. Loathe* is a stronger version of *dislike. Detest* and *hate,* Choices *A* and *B,* are more synonymous with *loathe* (rather than a different intensity), so they are not the best choice. Choice *C* is an antonym.

10. D: This is a tool/user analogy. The connection is the type of tool the artist uses to hold and create their work. An easel holds the paper or canvas that a painter uses to create a painting much like a loom is the apparatus used to hold and weave yarns into a blanket or other tapestry.

11. B: To *reproach* means to blame, find fault, or severely criticize. It can also be defined as expressing significant disapproval. It is often used in the phrase "beyond reproach" as in, "her violin performance was beyond reproach." In this context, it means her playing was so good that it evaded any possibility of criticism.

12. D: *Milieu* refers to the surroundings or environment.

13. A: *Guile* can be defined as the quality of being cunning or crafty and skilled in deception. Someone may use guile to trick or deceive someone, like to get money from them, or otherwise dupe them.

14. A: As a verb, *assent* means to express agreement, give consent, or acquiesce. A job candidate might assent to an interviewer's request to perform a background check. As a noun, it means an agreement, acceptance, or acquiescence.

15. B: *Dearth* means a lack of something or a scarcity or shortage. For example, a local library might have a dearth of information pertaining to an esoteric topic.

16. B: *Conspicuous* means to be visually or mentally obvious. Something conspicuous stands out, is clearly visible, or may attract attention.

17. D: *Onerous* most closely means burdensome or troublesome. Usually it is used to describe a task or obligation that may impose a hardship or burden, often which may be perceived to outweigh its benefits.

18: A: Something *banal* lacks originality, and may be boring and trite. For example, a banal compliment is likely to be a common platitude. Like something that is inane, a banal compliment might be meaningless and lack a convincing quality or significance.

19. C: *Contrite* means to feel or express remorse, or to be regretful and interested in repenting. The noun *contrition* refers to severe remorse or penitence.

20. A: *Mollify* means to soothe, pacify, or appease. It usually is used to refer to reducing the anger or softening the feelings or temper of another person, or otherwise calm them down. For example, a customer service associate may need to mollify an irate customer who is furious about the defect in his or her purchase.

21. D: Of the provided choices, *fickle* is closest in meaning to *capricious.* A person who is capricious tends to display erratic or unpredictable behavior, which is similar to fickle, which is also likely to change spontaneously or behave erratically.

22. C: Although *paltry* often is used as an adjective to describe a very small or meager amount (of money, in particular), it can also mean something trivial or insignificant.

23. B: The word *shirk* means to evade and is often used in the context of shirking a responsibility, duty, or work.

24. C: *Assuage* most nearly means to soothe or comfort, as in to assuage one's fears. It can also mean to lessen or make less severe, or to relieve. For example, an ice pack on a swollen knee may assuage the pain.

25. A: Something that is *tacit* is usually unspoken but implied. Tacit approval, for example, occurs when agreement or approval is understood without explicitly stating it.

26. B: To determine the answer, we can actually assign values to these items, which will help translate the information into practical terms.

Bagels:	$1.00
Croissants (less than bagels):	$0.75
Oatmeal (more than bagels):	$2.00

Once we assign arbitrary, but sound, numbers to these items, it can be seen that it is false that oatmeal costs less than croissants.

27. B: We can draw this out. Amelia has more siblings than Jude, so we can place her name higher than his. Jude has more siblings than Kendra, so below Jude, we can put Kendra. We can clearly see that the third statement is false, because Amelia is our highest name; she can't be below Kendra.

28. C: We are unable to have a sure answer here. We know mysteries are better than romances and fantasies aren't as good as mysteries, but we are uncertain whether fantasies are better than romances given the first two statements. We don't have enough information in the first two statements.

29. B: If all soups at Mimi's diner contain chicken stock, which is an animal product, the third statement is false because *all* soups—not just some—at Mimi's diner contain animal products.

30. C: This one is uncertain. We cannot say whether or not no runners are overweight. We only know that Ben is a runner and he (presumably one of many) is not overweight. Whether one, ten, or every other runner besides Ben is overweight is something we can't be sure of with the information we are given.

31. A: This one can be diagrammed if it is confusing. Lollipops are sweeter than jolly ranchers, so draw an L to the right of a J. Candy buttons are sweeter than lollipops, so draw a C even further to the right than the L. Now look and see that the J, jolly ranchers, are not as sweet (so they are to the left) as candy buttons, our furthest to the right, or sweetest, candy. The third statement is true.

32. B: We can start with the second statement here. All socks in the bin have cotton. Then we also know some socks have wool; thus, the third statement must be false: No socks are wool and cotton. Some socks *must* be wool and cotton since all socks have cotton and some socks in the bin have wool.

33. A: Whether or not you agree with the statements made here, according to the logic of the argument, the third statement is true. If no made-for-TV movies are entertaining, and Lifetime movies are made-for-TV movies, then by default, Lifetime movies are not entertaining.

34. C: The third statement here is uncertain. The first two statements tell us that both Seth and George drive further than Bill, but we aren't told if their commutes are equally long or if one drives further than the other.

35. A: This question is pretty straightforward. If some cars are considered to be eco-friendly vehicles and all eco-friendly vehicles are eligible to receive a rebate from the EPA, then some cars (the eco-friendly ones) are eligible to receive a rebate from the EPA. Therefore, the third statement is true.

36. A: The unifying attribute of three of the four of the words is that they are acids. *Bleach*, on the other hand, is a base, so it does not belong.

37. C: Trout, cod, and flounder are fish while shrimp is not.

38. B: A *sphere* is a solid, three-dimensional figure rather than a polygon like the other three shapes.

39. B: All of the terms are grammatically classified as adjectives except *nervously*, which is an adverb. Nervous would be the adjective analog.

40. D: Iron, Stone, and Ice are all named, notable ages in history. There was no Platinum Age.

41. C: In the answer choices provided, three of the words are relatively synonymous in meaning—*urgent, dire,* and *exigent* all refer to something that needs to be dealt with right away—and one word is unrelated. *Prudent* means practical or wise. Therefore, the correct choice is Choice *C, prudent*; it does not fit with the other words.

42. D: Three of the four professions listed work in the mental health field. A *physical therapist*, on the other hand, works with physical needs and deficits more so than psychological ones.

43. A: This one was a bit trickier. Discerning test takers would think beyond the more obvious definition of *table* to answer this question correctly. *Defer, table,* and *postpone* all relate to putting something off until a later time. *Schedule*, while somewhat related, is not synonymous with the other three terms.

44. C: *Utilize* is the "odd one out" because the other three words have to do with thinking things over or considering them.

45. C: At first glance, it might not be obvious what the overarching theme is to unify the words. Therefore, the word that doesn't belong might be tricky to spot. *Ate, discovered,* and *kicked* are verbs in the past tense, but *cook* is in present tense. The category or classification is grammatical in nature.

46. D: *Sinus* is not part of the eye whereas the other three terms are.

47. B: The unifying element of these terms is that they rhyme. *Carry* does not rhyme with the other three words, so it does not belong.

48. D: *Capitalizing*, which means using or taking advantage of something, is not synonymous with the other three terms, which all have to do with something being an object of interest—something people are drawn to.

49. B: *Dim, dull,* and *faint* are synonyms. *Spot* does not belong. It carries a different meaning.

50. B: This one was a bit trickier. *Rifle, scour,* and *rummage* are words that can pertain to searching through or examining something. *Concinnate* carries a different meaning; it means to assemble separate parts into a harmonious manner.

51. C: *Sheep* does not belong because the other three terms are baby animals, whereas a *sheep* is the adult form.

52. B: Someone who is *wary* is overly cautious or apprehensive. This word is often used in the context of being watchful or on guard about a potential danger. For example, darkening clouds and white caps on the waves may make a seaman wary against setting sail. Therefore, an antonym would be *eager.*

53. D: As an adjective, *parsimonious* means frugal to the point of being stingy, or very unwilling to spend money, which is similar to being miserly. Therefore, *generous* is a decent antonym.

54. B: The noun *propriety* means suitability or appropriateness to the given circumstances or purpose. It can mean conformity to accepted standards, particularly as they relate to good behavior or manners. In this way, *propriety* can be considered to mean the state or quality of being proper. Therefore, Choice *B,* inappropriateness, is the correct answer, because it is an antonym for propriety.

55. D: A *boon* is a benefit or blessing, often considered to be timely. It is something to be thankful for. For example, a new tax benefit enacted for first-time homeowners would be a boon to a family who just closed on a house. In an alternative usage, it can be a favor or a benefit given upon request. Therefore, *misfortune* is the best antonym.

56. B: *Strife* is a noun that is defined as bitter or vigorous discord, conflict, or dissension. It can mean a fight or struggle, or other act of contention. For example, antagonistic political interest groups vying for local support may be at strife. *Accord* means to be in harmony with, or agree. Therefore, *accord,* Choice *B,* is the antonym for *strife.*

57. A: *Qualm* is a noun that means a feeling of apprehension or uneasiness, often brought on suddenly. A girl who is just learning to ride a bike may have qualms about getting back on the saddle after taking a bad fall. It may also refer to an uneasy feeling related to one's conscience as it pertains to his or her actions. For example, a man with poor morals may have no qualms about lying on his tax return. *Certainty* is the best antonym in the given options.

58. D: *Valor* is bravery or courage when facing a formidable danger. It often relates to strength of mind or spirit during battle or acting heroically in such situations. Thus, *cowardice* is the best antonym.

59. B: *Zeal* is eagerness, fervor, or ardent desire in the pursuit of something. For example, a competitive collegiate baseball player's zeal to succeed in his sport may compromise his academic performance. *Apathy,* which means a lack of caring, is a good antonym.

60: B: To *extol* is to highly praise, laud, or glorify. People often extol the achievements of their heroes or mentors. *Castigate,* which means to berate or chastise, is an antonym.

Quantitative Skills

Number Series

Series and Sequences

A **sequence of numbers** is a list of numbers that follows a specific pattern. Each member of the sequence is known as an **individual term** of the sequence, and a formula can be found to represent each term. For example, the list of numbers 5, 10, 15, 20, ... is a sequence of numbers, and ... shows that the sequence continues indefinitely. Each term represents a multiple of 5. The first term is 1×5, the second term is 2×5, the third term is 3×5, etc. In general, the nth term is $5 \times n$. Other, more complicated sequences can exist as well. For example, each term of the sequence 1, 4, 9, 16, 25, ... is not found through addition or multiplication. Each term happens to be a perfect square, and the first term is 1 squared, the second term is 2 squared, etc. Therefore, the nth term is n^2. A famous sequence of numbers is the Fibonacci sequence, which consists of 0, 1, 1, 2, 3, 5, 8, 13, 21, ... After the first two terms 0 and 1, all terms are found by adding the two previous terms. Therefore, the next term in the sequence is $13 + 21 = 34$. The formula for the nth term is defined recursively, using the two previous terms.

Number and Shape Patterns

Patterns in math are those sets of numbers or shapes that follow a rule. Given a set of values, patterns allow the question of "what's next?" to be answered. In the following set, there are two types of shapes, a white rectangle and a gray circle. The set contains a pattern because every odd-placed shape is a white rectangle and every even-placed spot is taken by a gray circle. This is a pattern because there is a rule of white rectangle, then gray circle, that is followed to find the set.

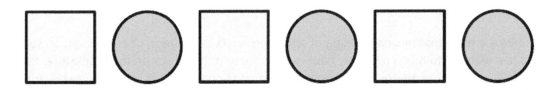

A set of numbers can also be described as having a pattern if there is a rule that can be followed to reproduce the set. The following set of numbers has a rule of adding 3 each time. It begins with zero and increases by 3 each time. By following this rule and pattern, the number after 12 is found to be 15. Further extending the pattern, the numbers are 18, 21, 24, 27. The pattern of increasing by multiples of three can describe this pattern.

$$0 \quad 3 \quad 6 \quad 9 \quad 12 \quad \boxed{} \quad ...$$

A pattern can also be generated from a given rule. Starting with zero, the rule of adding 5 can be used to produce a set of numbers. The following list will result from using the rule: 0, 5, 10, 15, 20. Describing this pattern can include words such as "multiples" of 5 and an "increase" of 5. Any time this pattern needs to be extended, the rule can be applied to find more numbers. Patterns are identified by the rules they follow. This rule should be able to generate new numbers or shapes, while also applying to the given numbers or shapes.

Making Predictions Based on Patterns

Given a certain pattern, future numbers or shapes can be found. **Pascal's triangle** is an example of a pattern of numbers. Questions can be asked of the triangle, such as, "what comes next?" and "what values determine the next line?" By examining the different parts of the triangle, conjectures can be made about how the numbers are generated. For the first few rows of numbers, the increase is small. Then the numbers begin to increase more quickly. By looking at each row, a conjecture can be made that the sum of the first row determines the second row's numbers. The second row's numbers can be added to find the third row.

To test this conjecture, two numbers can be added, and the number found directly between and below them should be that sum. For the third row, the middle number is 2, which is the sum of the two 1s above it. For the fifth row, the 1 and 3 can be added to find a sum of 4, the same number below the 1 and 3. This pattern continues throughout the triangle. Once the pattern is confirmed to apply throughout the triangle, predictions can be made for future rows. The sums of the bottom row numbers can be found and then added to the bottom of the triangle. In more general terms, the diagonal rows have patterns as well. The outside numbers are always 1. The second diagonal rows are in counting order. The third diagonal row increases each time by one more than the previous. It is helpful to generalize patterns because it makes the pattern more useful in terms of applying it. Pascal's triangle can be used to predict the tossing of a coin, predicting the chances of heads or tails for different numbers of tosses.

It can also be used to show the Fibonacci Sequence, which is found by adding the diagonal numbers together.

Pascal's Triangle

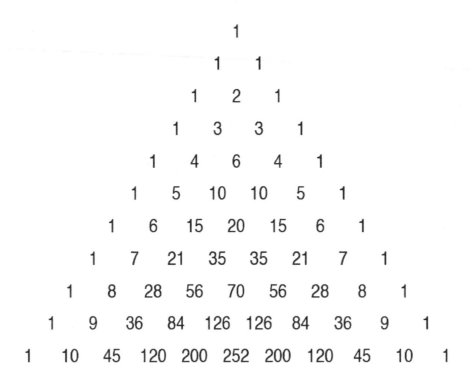

```
                        1
                     1     1
                  1     2     1
               1     3     3     1
            1     4     6     4     1
         1     5    10    10     5     1
      1     6    15    20    15     6     1
   1     7    21    35    35    21     7     1
1     8    28    56    70    56    28     8     1
1     9    36    84   126   126    84    36     9     1
1    10    45   120   200   252   200   120    45    10     1
```

Relationships Between the Corresponding Terms of Two Numerical Patterns

Sets of **numerical patterns** can be found by starting with a number and following a given rule. If two sets are generated, the corresponding terms in each set can be found to relate to one another by one or more operations. For example, the following table shows two sets of numbers that each follow their own pattern. The first column shows a pattern of numbers increasing by 1. The second column shows the numbers increasing by 4. Because the numbers are lined up, corresponding numbers are side by side for the two sets. A question to ask is, "How can the number in the first column be turned into the number in the second column?"

1	4
2	8
3	12
4	16
5	20

This answer will lead to the relationship between the two sets. By recognizing the multiples of 4 in the right column and the counting numbers in order in the left column, the relationship of multiplying by four is determined. The first set is multiplied by 4 to get the second set of numbers. To confirm this relationship, each set of corresponding numbers can be checked. For any two sets of numerical patterns, the corresponding numbers can be lined up to find how each one relates to the other. In some cases, the relationship is simply addition or subtraction, multiplication or division. In other relationships, these operations are used in conjunction with each together. As seen in the following table, the relationship uses multiplication and addition. The following expression shows this relationship: $3x + 2$. The x represents the numbers in the first column.

1	5
2	8
3	11
4	14

Geometric Comparison

Geometric Measurement and Dimension

Lines
Geometric figures can be identified by matching the definition with the object. For example, a **line segment** is made up of two endpoints and the line drawn between them. A **ray** is made up of one endpoint and one extending side that goes on forever. A **line** has no endpoints and two sides that extend on forever. These three geometric figures are shown below. What happens at A and B determines the name of each figure.

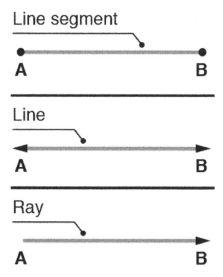

Parallel and perpendicular lines are made up of two lines, like the second figure above. They are distinguished from each other by how the two lines interact. **Parallel lines** run alongside one another, but they never intersect. **Perpendicular lines** intersect at a 90-degree, or a right, angle. An example of these two sets of lines is shown below. Also shown in the figure are non-examples of these two types of lines. Because the first set of lines, in the top left corner, will eventually intersect if they continue, they are not parallel. In the second set, the lines run in the same direction and will never intersect, making them parallel. The third set, in the bottom left corner, intersect at an angle that is not right, or not 90 degrees. The fourth set is perpendicular because it intersects at exactly a right angle.

Lines

Not Parallel	Parallel

Not Perpendicular	Perpendicular

Angles

When two rays are joined together at their endpoints, an **angle** is formed. Angles can be described based on their measure. An angle whose measure is 90 degrees is described as a right angle, just as with perpendicular lines. Ninety degrees is a standard, to which other angles are compared. If an angle is less than ninety degrees, it is an **acute angle**. If it is greater than ninety degrees, it is an **obtuse angle**. If an angle is equal to twice a right angle, or 180 degrees, it is a **straight angle**.

Examples of these types of angles are shown below:

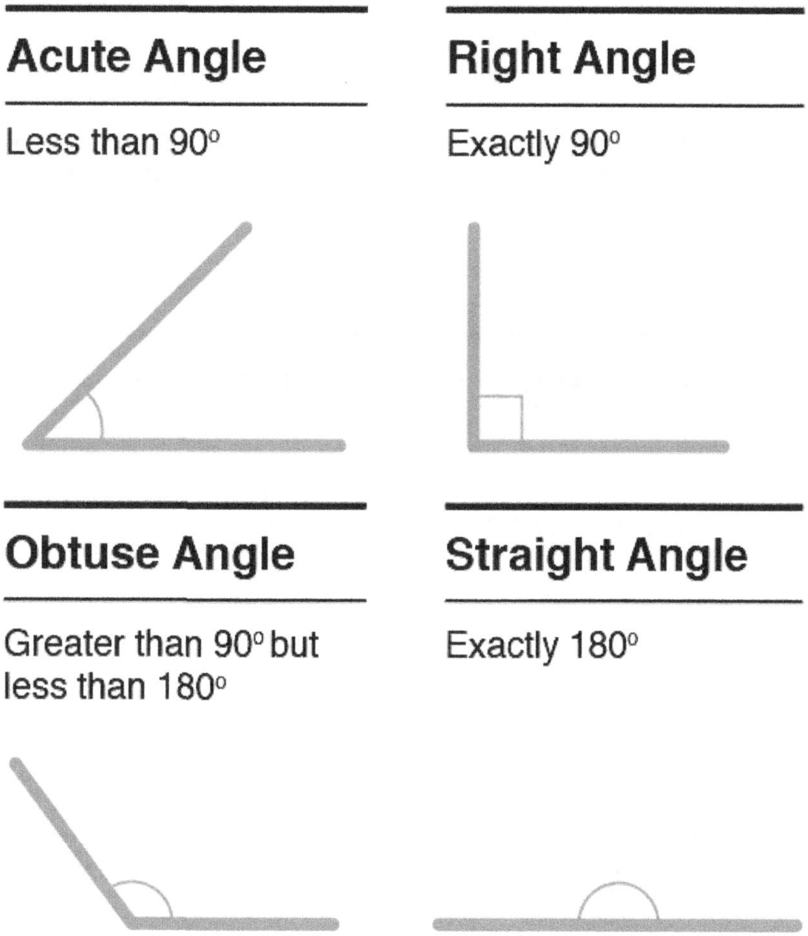

Acute Angle

Less than 90°

Right Angle

Exactly 90°

Obtuse Angle

Greater than 90° but less than 180°

Straight Angle

Exactly 180°

A **straight angle** is equal to 180 degrees, or a straight line. If the line continues through the **vertex**, or point where the rays meet, and does not change direction, then the angle is straight. This is shown in Figure 1 below. The second figure shows an obtuse angle. Its measure is greater than ninety degrees, but less than that of a straight angle. An estimate for its measure may be 175 degrees. Figure 3 shows an acute angle because it is just less than that of a right angle. Its measure may be estimated to be 80 degrees.

The last image, Figure 4, shows another acute angle. This measure is much smaller, at approximately 35 degrees, but it is still classified as acute because it is between zero and 90 degrees.

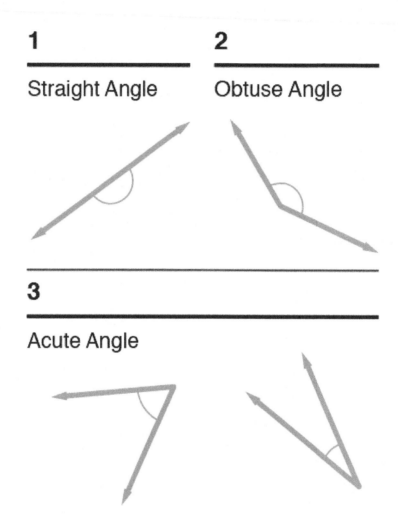

1

Straight Angle

2

Obtuse Angle

3

Acute Angle

Perimeter

Perimeter and area are two commonly used geometric quantities that describe objects. **Perimeter** is the distance around an object. The perimeter of an object can be found by adding the lengths of all sides. Perimeter may be used in problems dealing with lengths around objects such as fences or borders. It may also be used in finding missing lengths, or working backwards. If the perimeter is given, but a length is missing, subtraction can be used to find the missing length. Given a square with side length s, the formula for perimeter is $P = 4s$. Given a rectangle with length l and width w, the formula for perimeter is:

$$P = 2l + 2w$$

The perimeter of a triangle is found by adding the three side lengths, and the perimeter of a trapezoid is found by adding the four side lengths. The units for perimeter are always the original units of length, such as meters, inches, miles, etc. When discussing a circle, the distance around the object is referred to as its **circumference**, not perimeter. The formula for circumference of a circle is $C = 2\pi r$, where r represents the radius of the circle. This formula can also be written as $C = d\pi$, where d represents the diameter of the circle.

Area

Area is the two-dimensional space covered by an object. These problems may include the area of a rectangle, a yard, or a wall to be painted. Finding the area may be a simple formula, or it may require multiple formulas to be used together. The units for area are square units, such as square meters, square inches, and square miles. Given a square with side length s, the formula for its area is $A = s^2$. Some other common shapes are shown below:

Shape	Formula	Graphic
Rectangle	$Area = length \times width$	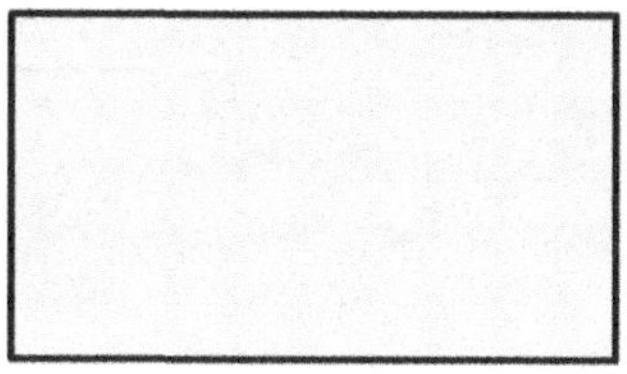
Triangle	$Area = \dfrac{1}{2} \times base \times height$	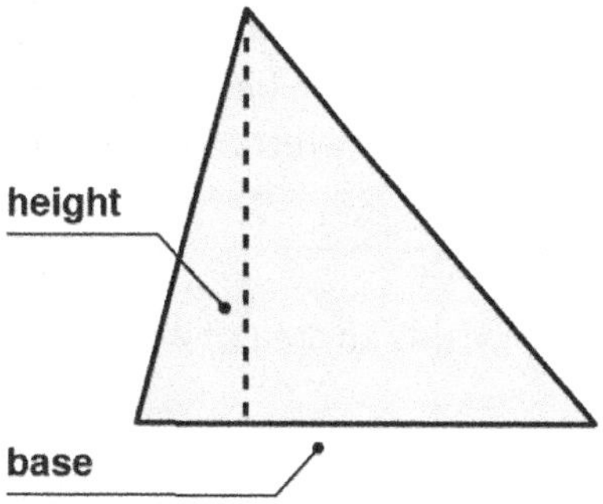
Circle	$Area = \pi \times radius^2$	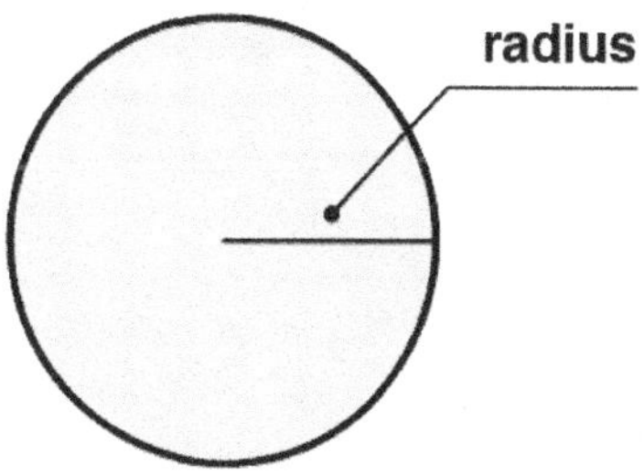

The following formula, not as widely used as those shown above, but very important, is the area of a trapezoid:

Area of a Trapezoid

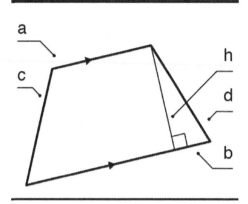

$$A = \frac{1}{2}\,(\,a + b\,)\,h$$

To find the area of the shapes above, use the given dimensions of the shape in the formula. Complex shapes might require more than one formula. To find the area of the figure below, break the figure into two shapes. The rectangle has dimensions 6 cm by 7 cm. The triangle has dimensions 6 cm by 6 cm. Plug the dimensions into the rectangle formula:

$$A = 6 \times 7$$

Multiplication yields an area of 42 cm². The triangle area can be found using the formula

$$A = \frac{1}{2} \times 4 \times 6$$

Multiplication yields an area of 12 cm². Add the areas of the two shapes to find the total area of the figure, which is 54 cm².

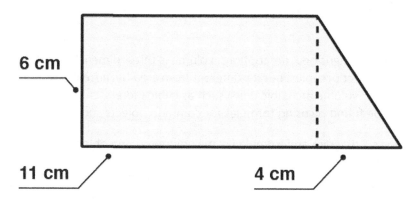

Instead of combining areas, some problems may require subtracting them, or finding the difference.

To find the area of the shaded region in the figure below, determine the area of the whole figure. Then the area of the circle can be subtracted from the whole.

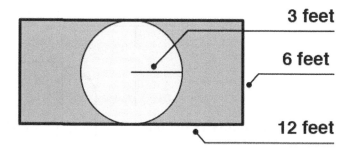

The following formula shows the area of the outside rectangle:

$$A = 12 \times 6 = 72 \text{ ft}^2$$

The area of the inside circle can be found by the following formula:

$$A = \pi(3)^2 = 9\pi = 28.3 \text{ ft}^2$$

As the shaded area is outside the circle, the area for the circle can be subtracted from the area of the rectangle to yield an area of 43.7 ft².

While some geometric figures may be given as pictures, others may be described in words. If a rectangular playing field with dimensions 95 meters long by 50 meters wide is measured for perimeter, the distance around the field must be found. The perimeter includes two lengths and two widths to measure the entire outside of the field. This quantity can be calculated using the following equation:

$$P = 2(95) + 2(50) = 290 \text{ m}$$

The distance around the field is 290 meters.

Volume

Perimeter and area are two-dimensional descriptions; volume is three-dimensional. **Volume** describes the amount of space that an object occupies, but it's different from area because it has three dimensions instead of two. The units for volume are cubic units, such as cubic meters, cubic inches, and cubic millimeters. Volume can be found by using formulas for common objects such as cylinders and boxes.

The following chart shows a diagram and formula for the volume of two objects:

Shape	Formula	Diagram
Rectangular Prism (box)	$V = length \times width \times height$	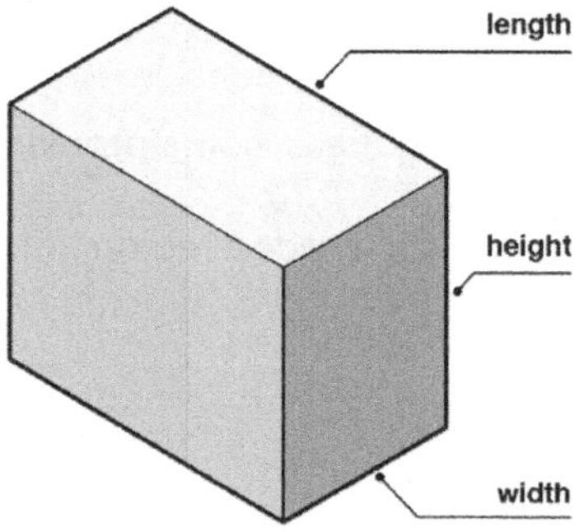
Cylinder	$V = \pi \times radius^2 \times height$	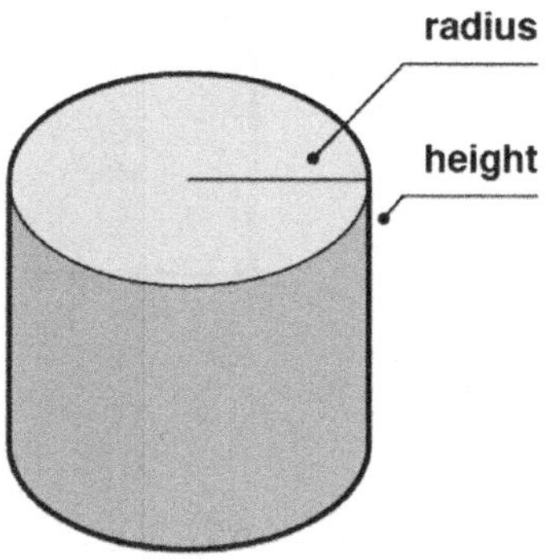

Volume formulas of these two objects are derived by finding the area of the bottom two-dimensional shape, such as the circle or rectangle, and then multiplying times the height of the three-dimensional shape. Other volume formulas include the volume of a cube with side length s: $V = s^3$; the volume of a sphere with radius r: $V = \frac{4}{3}\pi r^3$; and the volume of a cone with radius r and height h:

$$V = \frac{1}{3}\pi r^2 h$$

If a soda can has a height of 5 inches and a radius on the top of 1.5 inches, the volume can be found using one of the given formulas. A soda can is a cylinder. Knowing the given dimensions, the formula can be completed as follows:

$$V = \pi(radius)^2 \times height$$

$$\pi(1.5 \text{ in})^2 \times 5 \text{ in} = 35.325 \text{ in}^3$$

Notice that the units for volume are inches cubed because it refers to the number of cubic inches required to fill the can.

Right rectangular prisms are those prisms in which all sides are rectangles and all angles are right, or equal to 90 degrees. The volume for these objects can be found by multiplying the length by the width by the height. The formula is $V = lwh$. For the following prism, the volume formula is:

$$V = 6\frac{1}{2} \times 3 \times 9$$

When dealing with fractional edge lengths, it is helpful to convert the length to an improper fraction. The length $6\frac{1}{2}$ cm becomes $\frac{13}{2}$ cm. Then the formula becomes:

$$V = \frac{13}{2} \times 3 \times 9 = \frac{13}{2} \times \frac{3}{1} \times \frac{9}{1} = \frac{351}{2}$$

This value for volume is better understood when turned into a mixed number, which would be $175\frac{1}{2}$ cm^3.

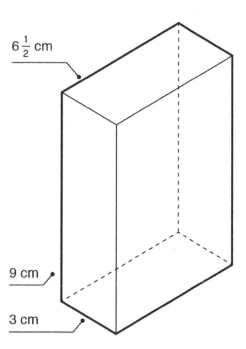

When dimensions for length are given with fractional parts, it can be helpful to turn the mixed number into an improper fraction, then multiply to find the volume, then convert back to a mixed number.

Surface Area

Surface area is defined as the area of the surface of a figure. A **pyramid** has a surface made up of four triangles and one square. To calculate the surface area of a pyramid, the areas of each individual shape are calculated. Then the areas are added together. This method of decomposing the shape into two-dimensional figures to find area, then adding the areas, can be used to find surface area for any figure. Once these measurements are found, the area is described with square units. For example, the following figure shows a rectangular prism. The figure beside it shows the rectangular prism broken down into two-dimensional shapes, or rectangles. The area of each rectangle can be calculated by multiplying the length by the width. The area for the six rectangles can be represented by the following expression:

$$5 \times 6 + 5 \times 10 + 5 \times 6 + 6 \times 10 + 5 \times 10 + 6 \times 10$$

The total for all these areas added together is 280 m², or 280 square meters. This measurement represents the surface area because it is the area of all six surfaces of the rectangular prism.

The Net of a Rectangular Prism

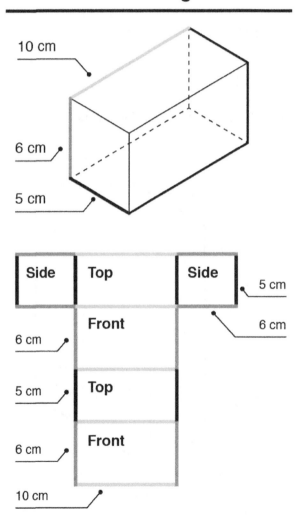

40

Another shape that has a surface area is a cylinder. The shapes used to make up the **cylinder** are two circles and a rectangle wrapped around between the two circles. A common example of a cylinder is a can. The two circles that make up the bases are obvious shapes. The rectangle can be more difficult to see, but the label on a can will help illustrate it. When the label is removed from a can and laid flat, the shape is a rectangle. When the areas for each shape are needed, there will be two formulas. The first is the area for the circles on the bases. This area is given by the formula $A = \pi r^2$. There will be two of these areas—one for the top and one for the bottom if the can (cylinder) is standing upright on a shelf. Then the area of the rectangle must be determined. The width of the rectangle is equal to the height of the can, h. The length of the rectangle is equal to the circumference of the base circle, $2\pi r$. The area for the rectangle can be found by using the formula $A = 2\pi r \times h$. By adding the two areas for the bases and the area of the rectangle, the surface area of the cylinder can be found, described in units squared.

The surface area of a rectangular prism can be found by breaking down the figure into basic shapes. These shapes are rectangles, made up of the two bases, two sides, and the front and back. Consider the rectangular prism we previously looked at:

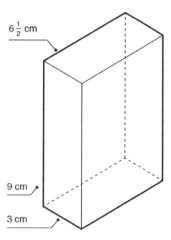

The formula for the surface area uses the area for each of these shapes for the terms in the following equation:

$$SA = 6\frac{1}{2} \times 3 + 6\frac{1}{2} \times 3 + 3 \times 9 + 3 \times 9 + 6\frac{1}{2} \times 9 + 6\frac{1}{2} \times 9$$

Because there are so many terms in a surface area formula and because this formula contains a fraction, it can be simplified by combining groups that are the same. Each set of numbers is used twice, to represent areas for the opposite sides of the prism. The formula can be simplified to

$$SA = 2\left(6\frac{1}{2} \times 3\right) + 2(3 \times 9) + 2\left(6\frac{1}{2} \times 9\right)$$

$$2\left(\frac{13}{2} \times 3\right) + 2(27) + 2\left(\frac{13}{2} \times 9\right)$$

$$2\left(\frac{39}{2}\right) + 54 + 2\left(\frac{117}{2}\right)$$

$$39 + 54 + 117 = 210 \text{ cm}^2$$

The **net of a figure** is the resulting two-dimensional shapes when a three-dimensional shape is broken down. Nets can be used in calculating different values for given shapes. One useful way to calculate surface area is to find the net of the object, then find the area of each of the shapes and add them together. Nets are also useful in composing shapes and decomposing objects so as to view how objects connect and can be used in conjunction with each other.

Surface area of three-dimensional figures is the total area of each of the faces of the figures. Nets are used to lay out each face of an object. The following figure shows a triangular prism. The bases are triangles and the sides are rectangles. The second figure shows the net for this triangular prism. The dimensions are labeled for each of the faces of the prism. The area for each of the two triangles can be determined by the formula:

$$A = \frac{1}{2}bh = \frac{1}{2} \times 8 \times 9 = 36 \text{ cm}^2$$

The rectangle areas can be described by the equation:

$$A = lw = 8 \times 5 + 9 \times 5 + 10 \times 5$$

$$40 + 45 + 50 = 135 \text{ cm}^2$$

The area for the triangles can be multiplied by two, then added to the rectangle areas to yield a total surface area of 207 cm².

A Triangular Prism and Its Net

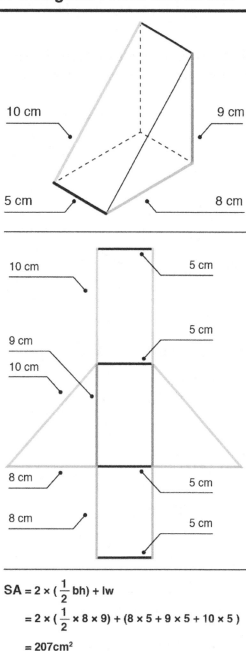

$$SA = 2 \times (\frac{1}{2}bh) + lw$$

$$= 2 \times (\frac{1}{2} \times 8 \times 9) + (8 \times 5 + 9 \times 5 + 10 \times 5)$$

$$= 207cm^2$$

Other figures that have rectangles or triangles in their nets include pyramids, rectangular prisms, and cylinders. When the shapes of these three-dimensional objects are found, and areas are calculated, the sum will result in the surface area. The following picture shows the net for a rectangular prism. The dimensions for each of the shapes that make up the prism are shown to the right. As a formula, the surface area is the sum of each shape added together. The following equation shows the formula:

$$SA = 5 \times 10 + 5 \times 6 + 6 \times 10 + 5 \times 6 + 5 \times 10 + 6 \times 10$$

$$SA = 50 + 30 + 60 + 30 + 50 + 60 = 280 \text{ m}^2$$

A Rectangular Prism and its Net

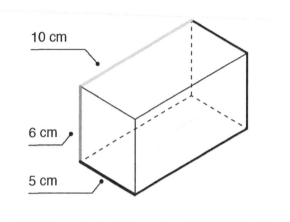

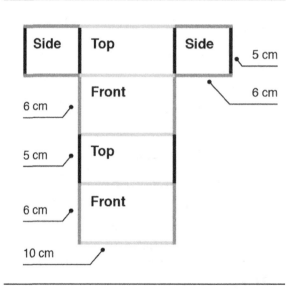

SA = 5×10 + 5×6 + 6×10 + 5×6 + 5×10 + 6×10

= 50 + 30 + 60 + 30 + 50 + 60

= 280cm²

A net for a pyramid is shown in the figure below. The base of the pyramid is a square. The four shapes coming off the pyramid are triangles. When built up together, folding the triangles to the top results in a pyramid. Surface area of the pyramid could be calculated by summing the area of each of the constituent shapes in the net.

The Net of a Pyramid

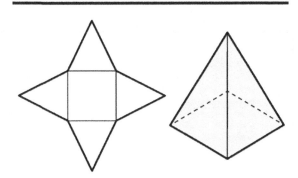

Another example of a net of a figure, in this case a cylinder, is shown below. When the cylinder is broken down, the bases are circles and the side is a rectangle wrapped around the circles. The circumference of the circle turns into the length of the rectangle.

The circumference
of the circle is
equal to the length
of the rectangle

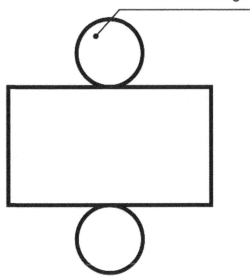

Determining How Changes to Dimensions Change Area and Volume
When the dimensions of an object change, the area and volume are also subject to change. For example, the following rectangle has an area of 98 square centimeters:

$$(A = 7 \times 14 = 98 \text{ cm}^2)$$

If the length is increased by 2, to be 16 cm, then the area becomes:

$$A = 7 \times 16 = 112 \text{ cm}^2$$

The area increased by 14 cm, or twice the width because there were two more widths of 7 cm.

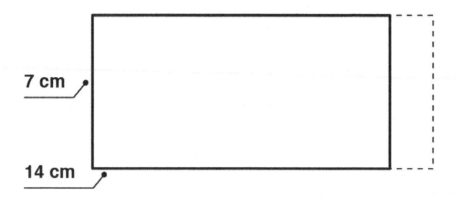

For the volume of an object, there are three dimensions. The given prism has a volume of:

$$V = 4 \times 12 \times 3 = 144 \text{ m}^3$$

If the height is increased by 3, the volume becomes:

$$V = 4 \times 12 \times 6 = 288 \text{ m}^3$$

The increase of 3 for the height, or doubling of the height, resulted in a volume that was doubled. From the original, if the width was doubled, the volume would be:

$$V = 8 \times 12 \times 3 = 288 \text{ m}^3$$

When the width doubled, the volume was doubled also. The same increase in volume would result if the length was doubled.

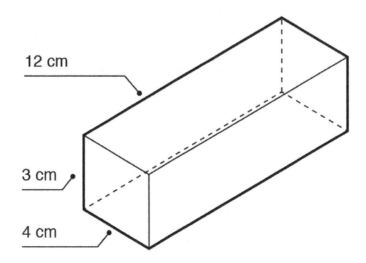

Solving for Missing Values in Triangles, Circles, and Other Figures

Solving for missing values in shapes requires knowledge of the shape and its characteristics. For example, a triangle has three sides and three angles that add up to 180 degrees. If two angle measurements are given, the third can be calculated. For the triangle below, the one given angle has a measure of 55 degrees. The missing angle is x. The third angle is labeled with a square, which indicates a measure of 90 degrees. Because all angles must sum to 180 degrees, the following equation can be used to find the missing x-value:

$$55° + 90° + x = 180°$$

Adding the two given angles and subtracting the total from 180, the missing angle is found to be 35 degrees.

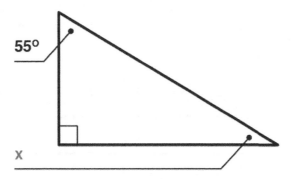

A similar problem can be solved with circles. If the radius is given but the circumference is unknown, the circumference can be calculated based on the formula $C = 2\pi r$. This example can be used in the figure below. The radius can be substituted for r in the formula. Then the circumference can be found as:

$$C = 2\pi \times 8 = 16\pi = 50.24 \text{ cm}$$

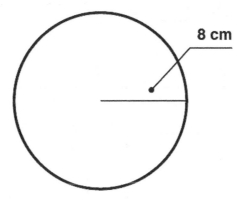

Other figures that may have missing values could be the length of a square, given the area, or the perimeter of a rectangle, given the length and width. All of the missing values can be found by first identifying all the characteristics that are known about the shape, then looking for ways to connect the missing value to the given information.

With any geometric calculations, it's important to determine what dimensions are given and what quantities the problem is asking for. If a connection can be made between them, the answer can be found.

Shapes and Solids

Shapes are defined by their angles and number of sides. A shape with one continuous side, where all points on that side are equidistant from a center point is called a **circle.** A shape made with three straight line segments is a **triangle.** A shape with four sides is called a **quadrilateral,** but more specifically a square, rectangle, parallelogram, or trapezoid, depending on the interior angles. These shapes are two-dimensional and only made of straight lines and angles.

Solids can be formed by combining these shapes and forming three-dimensional figures. These figures have another dimension because they add one more direction. Examples of solids may be prisms or spheres. There are four figures below that can be described based on their sides and dimensions. Figure 1 is a cone because it has three dimensions, where the bottom is a circle and the top is formed by the sides combining to one point. Figure 2 is a triangle because it has two dimensions, made up of three line segments. Figure 3 is a cylinder made up of two base circles and a rectangle to connect them in three dimensions. Figure 4 is an oval because it is one continuous line in two dimensions, not equidistant from the center.

Shapes and Solids

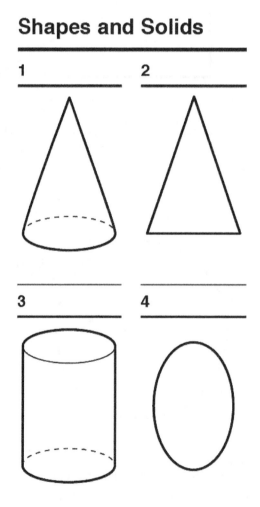

1

2

3

4

Figure 5 below is made up of squares in three dimensions, combined to make a cube. Figure 6 is a rectangle because it has four sides that intersect at right angles. More specifically, it can be described as a square because the four sides have equal measures. Figure 7 is a pyramid because the bottom shape is a

48

square and the sides are all triangles. These triangles intersect at a point above the square. Figure 8 is a circle because it is made up of one continuous line where the points are all equidistant from one center point.

Shapes and Solids

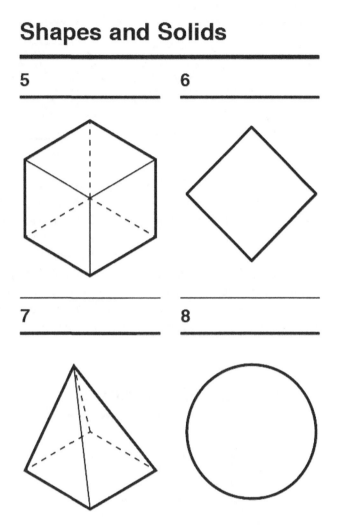

5

6

7

8

Decomposing Shapes

Basic shapes are those polygons that are made up of straight lines and angles and can be described by their number of sides and concavity. Some examples of those shapes are rectangles, triangles, hexagons, and pentagons. These shapes have identifying characteristics on their own, but they can also be decomposed into other shapes. For example, the following can be described as one hexagon, as seen in the first figure. It can also be decomposed into six equilateral triangles. The last figure shows how the hexagon can be decomposed into three rhombuses.

Decomposing a Hexagon

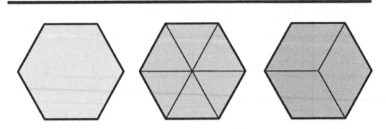

More complex shapes can be formed by combining basic shapes, or lining them up side by side. Below is an example of a house. This house is one figure all together, but can be decomposed into seven different shapes. The chimney is a parallelogram and the roof is made up of two triangles. The bottom of the house is a square alongside three triangles. There are many other ways of decomposing this house. Different shapes can be used to line up together and form one larger shape. The area for the house can be calculated by finding the individual areas for each shape, then adding them all together. For this house, there would be the area of four triangles, one square, and one parallelogram. Adding these all together would result in the area of the house as a whole. Decomposing and composing shapes is commonly done with a set of tangrams. A **tangram** is a set of shapes that includes different size triangles, rectangles, and parallelograms.

A Tangram of a House

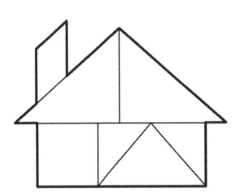

Congruence and Similarity

Two figures are **congruent** if they have the same shape and same size, meaning same angle measurements and equal side lengths. Two figures are **similar** if they have the same angle measurement but not side lengths. Basically, angles are congruent in similar triangles, and their side lengths are constant multiples of each other. Proving two shapes are similar involves showing that all angles are the same; proving two shapes are congruent involves showing that all angles are the same *and* that all sides are the same. If two pairs of angles are congruent in two triangles, then those triangles are similar because their third angle has to be equal due to the fact that all three angles add up to 180 degrees.

There are five main theorems that are used to show triangles are congruent. Each theorem involves showing different combinations of sides and angles are the same in two triangles, which proves the triangles are congruent. The **side-side-side (SSS) theorem** states that if all sides are equal in two triangles, the triangles are congruent. The **side-angle-side (SAS) theorem** states that if two pairs of sides are equal and the included angles are congruent in two triangles, then the triangles are congruent. Similarly, the **angle-side-angle (ASA) theorem** states that if two pairs of angles are congruent and the included side lengths are equal in two triangles, the triangles are similar. The **angle-angle-side (AAS) theorem** states that two triangles are congruent if they have two pairs of congruent angles and a pair of corresponding equal side lengths that are not included. Finally, the **hypotenuse-leg (HL) theorem** states that if two right triangles have equal hypotenuses and an equal pair of shorter sides, the triangles are congruent. An important item to note is that **angle-angle-angle (AAA)** is not enough information to have congruence because if three angles are equal in two triangles, the triangles can only be described as similar.

The Pythagorean Theorem and Right Triangles
Within right triangles, trigonometric ratios can be defined for the acute angle within the triangle. Consider the following right triangle. The side across from the right angle is known as the **hypotenuse**, the acute angle being discussed is labeled **θ**, the side across from the acute angle is known as the **opposite side**, and the other side is known as the **adjacent side**.

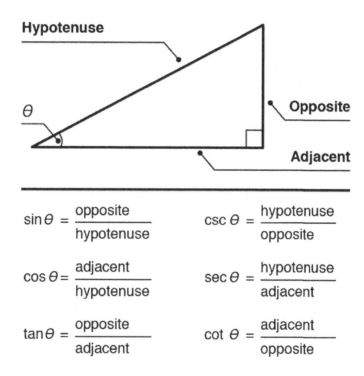

$$\sin \theta = \frac{\text{opposite}}{\text{hypotenuse}} \qquad \csc \theta = \frac{\text{hypotenuse}}{\text{opposite}}$$

$$\cos \theta = \frac{\text{adjacent}}{\text{hypotenuse}} \qquad \sec \theta = \frac{\text{hypotenuse}}{\text{adjacent}}$$

$$\tan \theta = \frac{\text{opposite}}{\text{adjacent}} \qquad \cot \theta = \frac{\text{adjacent}}{\text{opposite}}$$

The six trigonometric ratios are shown above as well. "**Sin**" is short for sine, "**cos**" is short for cosine, "**tan**" is short for tangent, "**csc**" is short for cosecant, "**sec**" is short for secant, and "**cot**" is short for cotangent. A mnemonic device exists that is helpful to remember the ratios. **SOHCAHTOA** stands for

Sine = Opposite/Hypotenuse, Cosine = Adjacent/Hypotenuse, and Tangent = Opposite/Adjacent. The other three trigonometric ratios are reciprocals of sine, cosine, and tangent because:

$$\csc\theta = \frac{1}{\sin\theta}$$

$$\sec\theta = \frac{1}{\cos\theta}$$

and

$$\cot\theta = \frac{1}{\tan\theta}$$

The **Pythagorean Theorem** is an important relationship between the three sides of a right triangle. It states that the square of hypotenuse is equal to the sum of the squares of the other two sides. When using the Pythagorean Theorem, the hypotenuse is labeled as side c, the opposite is labeled as side a, and the adjacent side is side b.

The theorem can be seen in the following diagram:

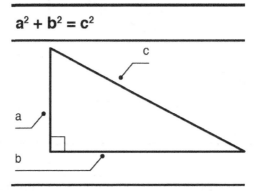

Both the trigonometric ratios and Pythagorean Theorem can be used in problems that involve finding either a missing side or missing angle of a right triangle. Look to see what sides and angles are given and select the correct relationship that will assist in finding the missing value. These relationships can also be used to solve application problems involving right triangles. Often, it is helpful to draw a figure to represent the problem to see what is missing.

Plane Geometry

The Coordinate Plane
The coordinate plane is a way of identifying the position of a point in relation to two axes. The **coordinate plane** is made up of two intersecting lines, the x-axis and the y-axis. These lines intersect at a right angle, and their intersection point is called the **origin**. The points on the coordinate plane are labeled based on their position in relation to the origin. If a point is found 4 units to the right and 2 units up from the origin, the location is described as (4, 2). These numbers are the **x- and y-coordinates**, always written in

the order (x, y). This point is also described as lying in the first quadrant. Every point in the first quadrant has a location that is positive in the x and y directions.

The following figure shows the coordinate plane with examples of points that lie in each quadrant.

The Coordinate Plane

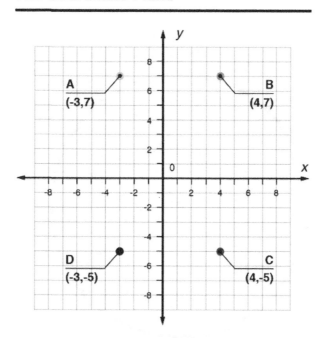

Point B lies in the first quadrant, described with positive x- and y-values, above the x-axis and to the right of the y-axis. Point A lies in the second quadrant, where the x-value is negative and the y-value is positive. This quadrant is above the x-axis and to the left of the y-axis. Point D lies in the third quadrant, where both the x- and y-values are negative. Points in this quadrant are described as being below the x-axis and to the left of the y-axis. Point C is in the fourth quadrant, where the x-value is positive and the y-value is negative.

Recall that a **circle** is the set of all points the same distance, known as **radius** r, from a single point C, known as the **center of the circle**. The center has coordinates (h, k) and any point on the circle is an ordered pair with coordinates (x, y). A right triangle with hypotenuse r can be formed with these two points as seen here:

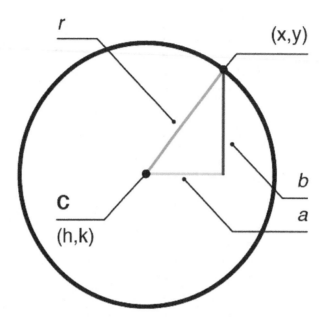

The other side lengths are a and b. The Pythagorean Theorem states that:

$$a^2 + b^2 = r^2$$

However, a can be replaced by $|x - h|$ and b can be replaced by $|y - k|$ because the distance between any two coordinates in the coordinate plane is the absolute value of their difference. That substitution gives:

$$(x - h)^2 + (y - k)^2 = r^2$$

which is the formula used to find the equation of any circle with center (h, k) and radius r. Therefore, if any problem gives the coordinates of the center of a circle and its radius length, this is the equation in two variables that allows any other point on the circle to be found.

Oftentimes, the center or the radius of a circle are not easily seen in the given equation of the circle. If the equation is in standard form of a polynomial equation like:

$$ax^2 + ay^2 + cx + dy + e = 0$$

the algebraic technique of completing the square must be used to find the coordinates of the center and the radius. Completing the square must be done within both variables x and y. First, the constant term needs to be subtracted off of both sides of the equation, and then the x and y terms need to be grouped together. Then, the entire equation needs to be divided by a. Then, divide the coefficient of the x term by

2, square it, and add that value to both sides of the equation. This value should be grouped with the x terms. Next, divide the coefficient of the y term by 2, square it, and add it to both sides of the equation, grouping it with the y terms. The trinomial in both x and y are now perfect square trinomials and can be factored into squares of a binomial. This process results in:

$$(x - h)^2 + (y - k)^2 = r^2$$

showing the radius and coordinates of the center.

Transformations in the Plane
Two-dimensional figures can undergo various types of transformations in the plane. They can be translated (shifted) horizontally and vertically, reflected, compressed, or stretched.

A **translation,** also known as a **shift** or **slide,** moves the shape in one direction. Here is a picture of a translation:

A Translation

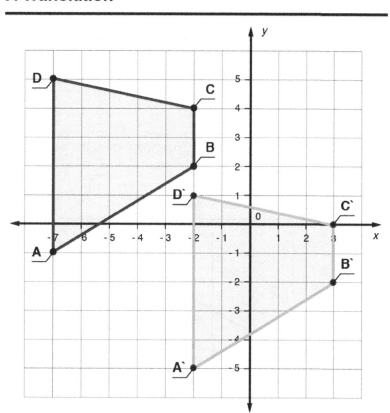

Notice that the size of the original shape has not changed at all. If a translation exists within a shape drawn on the Cartesian coordinate system, the translation can be represented by adding or subtracting values onto the x-, y-, or both the x- and y-coordinates of the points. However, all vertices will move the same number of units because the shape and size of the shape do not change.

A figure can also be reflected or flipped, and this transformation involves reflecting over a given line, known as the **line of reflection**. The original triangle (called the **preimage**) is seen in the figure below in

55

the first quadrant. The reflection of this triangle is in the second quadrant. A reflection across the y-axis can be found by determining each point's distance to the y-axis and moving it that same distance on the opposite side. For example, the point C is located at (4, 1). The reflection of this point moves to (-4, 1) when reflected across the y-axis. The original point A is located at (1, 3), and the reflection across the y-axis is located at (-1, 5). It is evident that the reflection across the y-axis changes the sign of the x-coordinate. A reflection across the x-axis changes the sign of the y-coordinate instead.

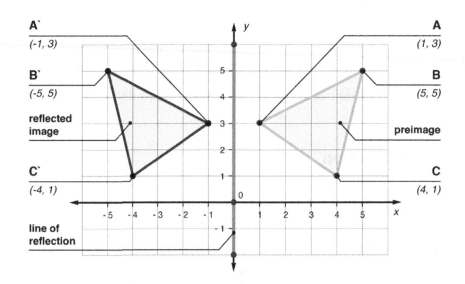

Similarly, if the shape is reflected over the x-axis, the y-coordinate stays the same, but the x-coordinates are made negative. For instance, in the graphic below, the point C at (3, 5) becomes C' at (3, -5).

A Reflection Over the X-Axis

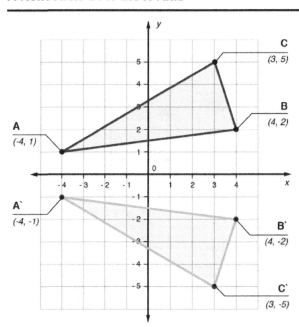

Thirdly, a **compression** or a stretch of a figure involves changing the size of the original figure, and together they can be classified as a **dilation**. A compression shrinks the size of the figure. We can think about this as a multiplication process by multiplying times a value between 0 and 1. A **stretch** of a figure results in a figure larger than the original shape. If we consider multiplication, the factor would be greater than 1. Here is a picture of a dilation that is comprised of a stretch in which the original square quadrupled in size after the length of each side increased by a factor of 2:

A Dilation with a Scale Factor of 2

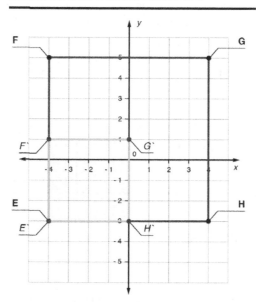

If a shape within the Cartesian coordinate system gets stretched, its coordinates get multiplied by a number greater than 1, and if a shape gets compressed, its coordinates get multiplied by a number between 0 and 1.

A figure can undergo any combination of transformations. For instance, it can be shifted, reflected, and stretched at the same time.

Converting Within and Between Standard and Metric Systems

When working with dimensions, sometimes the given units don't match the formula, and conversions must be made. The **metric system** has base units of meter for length, kilogram for mass, and liter for liquid volume. This system expands to three places above the base unit and three places below. These places correspond with prefixes with a base of 10.

The following table shows the conversions:

kilo-	hecto-	deka-	base	deci-	centi-	milli-
1,000 times the base	100 times the base	10 times the base		1/10 times the base	1/100 times the base	1/1000 times the base

To convert between units within the metric system, values with a base ten can be multiplied. The decimal can also be moved in the direction of the new unit by the same number of zeros on the number. For example, 3 meters is equivalent to 0.003 kilometers. The decimal moved three places (the same number of zeros for kilo-) to the left (the same direction from base to kilo-). Three meters is also equivalent to 3,000 millimeters. The decimal is moved three places to the right because the prefix milli- is three places to the right of the base unit.

The English Standard system used in the United States has a base unit of foot for length, pound for weight, and gallon for liquid volume. These conversions aren't as easy as the metric system because they aren't a base ten model. The following table shows the conversions within this system.

Length	Weight	Capacity
1 foot (ft) = 12 inches (in) 1 yard (yd) = 3 feet 1 mile (mi) = 5280 feet 1 mile = 1760 yards	1 pound (lb) = 16 ounces (oz) 1 ton = 2000 pounds	1 tablespoon (tbsp) = 3 teaspoons (tsp) 1 cup (c) = 16 tablespoons 1 cup = 8 fluid ounces (oz) 1 pint (pt) = 2 cups 1 quart (qt) = 2 pints 1 gallon (gal) = 4 quarts

When converting within the English Standard system, most calculations include a conversion to the base unit and then another to the desired unit. For example, take the following problem:

$$3 \text{ qt} = \underline{\quad} \text{ c}$$

There is no straight conversion from quarts to cups, so the first conversion is from quarts to pints. There are 2 pints in 1 quart, so there are 6 pints in 3 quarts. This conversion can be solved as a proportion:

$$\frac{3 \text{ qt}}{x} = \frac{1 \text{ qt}}{2 \text{ pt}}$$

It can also be observed as a ratio 2:1, expanded to 6:3. Then the 6 pints must be converted to cups. The ratio of pints to cups is 1:2, so the expanded ratio is 6:12. For 6 pints, the measurement is 12 cups. This problem can also be set up as one set of fractions to cancel out units. It begins with the given information and cancels out matching units on top and bottom to yield the answer. Consider the following expression:

$$\frac{3 \text{ qt}}{1} \times \frac{2 \text{ pt}}{1 \text{ qt}} \times \frac{2 \text{ c}}{1 \text{ pt}}$$

It's set up so that units on the top and bottom cancel each other out:

$$\frac{3 \text{ \cancel{qt}}}{1} \times \frac{2 \text{ \cancel{pt}}}{1 \text{ \cancel{qt}}} \times \frac{2 \text{ c}}{1 \text{ \cancel{pt}}}$$

The numbers can be calculated as $3 \times 2 \times 2$ on the top and 1 on the bottom. It still yields an answer of 12 cups.

This process of setting up fractions and canceling out matching units can be used to convert between standard and metric systems. A few common equivalent conversions are:

$$2.54 \text{ cm} = 1 \text{ in}$$

$$3.28 \text{ ft} = 1 \text{ m}$$

and

$$2.205 \text{ lb} = 1 \text{ kg}$$

Writing these as fractions allows them to be used in conversions. For the fill-in-the-blank problem 5 m = ___ ft, an expression using conversions starts with the expression:

$$\frac{5 \text{ m}}{1} \times \frac{3.28 \text{ ft}}{1 \text{ m}}$$

where the units of meters will cancel each other out and the final unit is feet. Calculating the numbers yields 16.4 feet. This problem only required two fractions. Others may require longer expressions, but the underlying rule stays the same. When there's a unit on the top of the fraction that's the same as the unit on the bottom, then they cancel each other out. Using this logic and the conversions given above, many units can be converted between and within the different systems.

The conversion between Fahrenheit and Celsius is found in a formula:

$$°C = (°F - 32) \times \frac{5}{9}$$

For example, to convert 78°F to Celsius, the given temperature would be entered into the formula:

$$°C = (78 - 32) \times \frac{5}{9}$$

Solving the equation, the temperature comes out to be 25.56°C. To convert in the other direction, the formula becomes:

$$°F = °C \times \frac{9}{5} + 32$$

Remember the order of operations when calculating these conversions.

Nongeometric Comparison

Solving Practical Math Problems

One-step problems take only one mathematical step to solve. For example, solving the equation $5x = 45$ is a one-step problem because the one step of dividing both sides of the equation by 5 is the only step necessary to obtain the solution $x = 9$. The **multiplication principle of equality** is the one step used to isolate the variable. The equation is of the form $ax = b$, where a and b are rational numbers. Similarly, the **addition principle of equality** could be the one step needed to solve a problem. In this case, the equation would be of the form $x + a = b$ or $x - a = b$, for real numbers a and b.

A **multi-step problem** involves more than one step to find the solution, or it could consist of solving more than one equation. An equation that involves both the addition principle and the multiplication principle is a two-step problem, and an example of such an equation is:

$$2x - 4 = 5$$

Solving involves adding 4 to both sides and then dividing both sides by 2. An example of a two-step problem involving two separate equations is:

$$y = 3x$$

$$2x + y = 4$$

The two equations form a system of two equations that must be solved together in two variables. The system can be solved by the substitution method. Since y is already solved for in terms of x, plug $3x$ in for y into the equation:

$$2x + y = 4$$

resulting in

$$2x + 3x = 4$$

Therefore:

$$5x = 4$$

and

$$x = \frac{4}{5}$$

Because there are two variables, the solution consists of a value for both x and for y. Substitute $x = \frac{4}{5}$ into either original equation to find y. The easiest choice is $y = 3x$. Therefore:

$$y = 3 \times \frac{4}{5} = \frac{12}{5}$$

The solution can be written as the ordered pair $\left(\frac{4}{5}, \frac{12}{5}\right)$.

Real-world problems can be translated into both one-step and multi-step problems. In either case, the word problem must be translated from the verbal form into mathematical expressions and equations that can be solved using algebra. An example of a one-step real-world problem is the following: A cat weighs half as much as a dog living in the same house. If the dog weighs 14.5 pounds, how much does the cat weigh? To solve this problem, an equation can be used. In any word problem, the first step must be defining variables that represent the unknown quantities. For this problem, let x be equal to the unknown weight of the cat. Because two times the weight of the cat equals 14.5 pounds, the equation to be solved is:

$$2x = 14.5$$

Use the multiplication principle to divide both sides by 2. Therefore, $x = 7.25$. The cat weighs 7.25 pounds.

Most of the time, real-world problems are more difficult than this one and are multi-step problems. The following is an example of a multi-step problem: The sum of two consecutive page numbers is equal to 437. What are those page numbers? First, define the unknown quantities. If x is equal to the first page number, then $x + 1$ is equal to the next page number because they are consecutive integers. Their sum is equal to 437, and this statement translates to the equation:

$$x + (x + 1) = 437$$

To solve, first collect like terms to obtain:

$$2x + 1 = 437$$

Then, subtract 1 from both sides and then divide by 2. The solution to the equation is $x = 218$. Therefore, the two consecutive page numbers that satisfy the problem are 218 and 219. It is always important to make sure that answers to real-world problems make sense. For instance, it should be a red flag if the solution to this same problem resulted in decimals, which would indicate the need to check the work. Page numbers are whole numbers; therefore, if decimals are found to be answers, the solution process should be double-checked to see where mistakes were made.

Solving Single- and Multistep Problems Involving Percentages
Percentages are defined to be parts per one hundred. To convert a decimal to a percentage, move the decimal point two units to the left and place the percent sign after the number. Percentages appear in many scenarios in the real world. It is important to make sure the statement containing the percentage is translated to a correct mathematical expression. Be aware that it is extremely common to make a mistake when working with percentages within word problems.

An example of a word problem containing a percentage is the following: 35% of people speed when driving to work. In a group of 5,600 commuters, how many would be expected to speed on the way to their place of employment? The answer to this problem is found by finding 35% of 5,600. First, change the percentage to the decimal 0.35. Then compute the product:

$$0.35 \times 5,600 = 1,960$$

Therefore, it would be expected that 1,960 of those commuters would speed on their way to work based on the data given. In this situation, the word "of" signals to use multiplication to find the answer. Another way percentage is used is in the following problem: _Teachers work 8 months out of the year. What percent of the year do they work?_ To answer this problem, find what percent _of_ 12 the number 8 _is,_ because there are 12 months in a year. Therefore, divide 8 by 12, and convert that number to a percentage:

$$\frac{8}{12} = \frac{2}{3} = 0.66\overline{6}$$

The percentage rounded to the nearest tenth place tells us that teachers work 66.7% of the year. Percentages also appear in real-world application problems involving finding missing quantities like in the following question: 60% of what number is 75? To find the missing quantity, an equation can be used. Let x be equal to the missing quantity. Therefore, $0.60x = 75$. Divide each side by 0.60 to obtain 125. Therefore, 60% of 125 is equal to 75.

Sales tax is an important application relating to percentages because tax rates are usually given as percentages. For example, a city might have an 8% sales tax rate. Therefore, when an item is purchased with that tax rate, the real cost to the customer is 1.08 times the price in the store. For example, a $25 pair of jeans costs the customer:

$$\$25 \times 1.08 = \$27$$

Sales tax rates can also be determined if they are unknown when an item is purchased. If a customer visits a store and purchases an item for $21.44, but the price in the store was $19, they can find the tax rate by first subtracting:

$$\$21.44 - \$19$$

to obtain $2.44, the sales tax amount. The sales tax is a percentage of the in-store price. Therefore, the tax rate is:

$$\frac{2.44}{19} = 0.128$$

which has been rounded to the nearest thousandths place. In this scenario, the actual sales tax rate given as a percentage is 12.8%.

Solving Real-World Problems Involving Proportions
Fractions appear in everyday situations, and in many scenarios, they appear in the real-world as ratios and in proportions. A **ratio** is formed when two different quantities are compared. For example, in a group of 50 people, if there are 33 females and 17 males, the ratio of females to males is 33 to 17. This expression can be written in the fraction form as $\frac{33}{50}$, where the denominator is the sum of females and males, or by using the ratio symbol, 33:17. The order of the number matters when forming ratios. In the same setting, the ratio of males to females is 17 to 33, which is equivalent to $\frac{17}{50}$ or 17:33.

A **proportion** is an equation involving two ratios. The equation:

$$\frac{a}{b} = \frac{c}{d}$$

or

$$a:b = c:d$$

is a proportion, for real numbers a, b, c, and d. Usually, in one ratio, one of the quantities is unknown, and cross-multiplication is used to solve for the unknown. Consider:

$$\frac{1}{4} = \frac{x}{5}$$

To solve for x, cross-multiply to obtain $5 = 4x$. Divide each side by 4 to obtain the solution:

$$x = \frac{5}{4}$$

It is also true that percentages are ratios in which the second term is 100 minus the first term. For example, 65% is 65:35 or $\frac{65}{100}$. Therefore, when working with percentages, one is also working with ratios.

Real-world problems frequently involve proportions. For example, consider the following problem: If 2 out of 50 pizzas are usually delivered late from a local Italian restaurant, how many would be late out of 235 orders? The following proportion would be solved with x as the unknown quantity of late pizzas:

$$\frac{2}{50} = \frac{x}{235}$$

Cross multiplying results in:

$$470 = 50x$$

Divide both sides by 50 to obtain $x = \frac{470}{50}$, which in lowest terms is equal to $\frac{47}{5}$. In decimal form, this improper fraction is equal to 9.4. Because it does not make sense to answer this question with decimals (portions of pizzas do not get delivered) the answer must be rounded. Traditional rounding rules would say that 9 pizzas would be expected to be delivered late. However, to be safe, rounding up to 10 pizzas out of 235 would probably make more sense.

Solving Real-World Problems Involving Ratios and Rates of Change
Recall that a ratio is the comparison of two different quantities. Comparing 2 apples to 3 oranges results in the ratio 2:3, which can be expressed as the fraction $\frac{2}{5}$. Note that order is important when discussing ratios. The number mentioned first is the antecedent, and the number mentioned second is the consequent. Note that the consequent of the ratio and the denominator of the fraction are *not* the same. When there are 2 apples to 3 oranges, there are five fruit total; two fifths of the fruit are apples, while three fifths are oranges. The ratio 2:3 represents a different relationship that the ratio 3:2. Also, it is important to make sure that when discussing ratios that have units attached to them, the two quantities use the same units. For example, to think of 8 feet to 4 yards, it would make sense to convert 4 yards to feet by multiplying by 3. Therefore, the ratio would be 8 feet to 12 feet, which can be expressed as the fraction $\frac{8}{20}$. Also, note that it is proper to refer to ratios in lowest terms. Therefore, the ratio of 8 feet to 4 yards is equivalent to the fraction $\frac{2}{5}$.

Many real-world problems involve ratios. Often, problems with ratios involve proportions, as when two ratios are set equal to find the missing amount. However, some problems involve deciphering single ratios. For example, consider an amusement park that sold 345 tickets last Saturday. If 145 tickets were sold to adults and the rest of the tickets were sold to children, what would the ratio of the number of adult tickets to children's tickets be? A common mistake would be to say the ratio is 145:345. However, 345 is the total number of tickets sold, not the number of children's tickets. There were:

$$345 - 145 = 200$$

tickets sold to children. The correct ratio of adult to children's tickets is 145:200. As a fraction, this expression is written as $\frac{145}{345}$, which can be reduced to $\frac{29}{69}$.

While a ratio compares two measurements using the same units, rates compare two measurements with different units. Examples of rates would be $200 for 8 hours of work, or 500 miles traveled per 20 gallons. Because the units are different, it is important to always include the units when discussing rates. Rates can be easily seen because if they are expressed in words, the two quantities are usually split up using one of the following words: *for, per, on, from, in.* Just as with ratios, it is important to write rates in lowest terms. A common rate that can be found in many real-life situations is cost per unit. This quantity describes how much one item or one unit costs. This rate allows the best buy to be determined, given a couple of different sizes of an item with different costs. For example, if 2 quarts of soup was sold for $3.50 and 3 quarts was sold for $4.60, to determine the best buy, the cost per quart should be found. $\frac{\$3.50}{2 \text{ qt}} = \$1.75$ per quart, and:

$$\frac{\$4.60}{3 \text{ qt}} = \$1.53 \text{ per quart}$$

Therefore, the better deal would be the 3-quart option.

Rate of change problems involve calculating a quantity per some unit of measurement. Usually the unit of measurement is time. For example, meters per second is a common rate of change. To calculate this measurement, find the amount traveled in meters and divide by total time traveled. The calculation is an average of the speed over the entire time interval. Another common rate of change used in the real world is miles per hour. Consider the following problem that involves calculating an average rate of change in temperature. Last Saturday, the temperature at 1:00 a.m. was 34 degrees Fahrenheit, and at noon, the temperature had increased to 75 degrees Fahrenheit. What was the average rate of change over that time interval? The average rate of change is calculated by finding the change in temperature and dividing by the total hours elapsed. Therefore, the rate of change was equal to:

$$\frac{75 - 34}{12 - 1} = \frac{41}{11}$$

degrees per hour. This quantity rounded to two decimal places is equal to 3.72 degrees per hour.

A common rate of change that appears in algebra is the slope calculation. Given a linear equation in one variable, $y = mx + b$, the **slope**, m, is equal to:

$$\frac{rise}{run}$$

or

$$\frac{change\ in\ y}{change\ in\ x}$$

In other words, slope is equivalent to the ratio of the vertical and horizontal changes between any two points on a line. The vertical change is known as the **rise**, and the horizontal change is known as the **run**. Given any two points on a line (x_1, y_1) and (x_2, y_2), slope can be calculated with the formula:

$$m = \frac{y_2 - y_1}{x_2 - x_1} = \frac{\Delta y}{\Delta x}$$

Common real-world applications of slope include determining how steep a staircase should be, calculating how steep a road is, and determining how to build a wheelchair ramp.

Many times, problems involving rates and ratios involve proportions. A proportion states that two ratios (or rates) are equal. The property of cross products can be used to determine if a proportion is true, meaning both ratios are equivalent. If $\frac{a}{b} = \frac{c}{d}$, then to clear the fractions, multiply both sides by the least common denominator, bd. This results in $ad = bc$, which is equal to the result of multiplying along both diagonals. For example:

$$\frac{4}{40} = \frac{1}{10}$$

grants the cross product:

$$4 \times 10 = 40 \times 1$$

which is equivalent to $40 = 40$ and shows that this proportion is true. Cross products are used when proportions are involved in real-world problems. Consider the following: If 3 pounds of fertilizer will cover 75 square feet of grass, how many pounds are needed for 375 square feet? To solve this problem, a proportion can be set up using two ratios. Let x equal the unknown quantity, pounds needed for 375 feet. Then, the equation found by setting the two given ratios equal to one another is:

$$\frac{3}{75} = \frac{x}{375}$$

Cross-multiplication gives:

$$3 \times 375 = 75x$$

Therefore, $1{,}125 = 75x$. Divide both sides by 75 to get $x = 15$. Therefore, 15 pounds of fertilizer are needed to cover 375 square feet of grass.

Another application of proportions involves similar triangles. If two triangles have the same measurement as two triangles in another triangle, the triangles are said to be **similar**. If two are the same, the third pair of angles are equal as well because the sum of all angles in a triangle is equal to 180 degrees. Each pair of equivalent angles are known as **corresponding angles. Corresponding sides** face the corresponding angles, and it is true that corresponding sides are in proportion. For example, consider the following set of similar triangles:

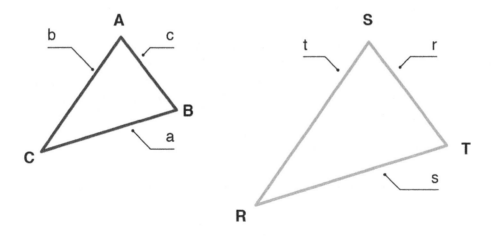

Angles A and S have the same measurement, angles C and R have the same measurement, and angles B and T have the same measurement. Therefore, the following proportion can be set up from the sides:

$$\frac{c}{r} = \frac{a}{s} = \frac{b}{t}$$

This proportion can be helpful in finding missing lengths in pairs of similar triangles. For example, if the following triangles are similar, a proportion can be used to find the missing side lengths, a and b.

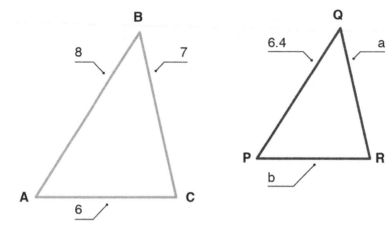

The proportions:

$$\frac{8}{6.4} = \frac{6}{b}$$

and

$$\frac{8}{6.4} = \frac{7}{a}$$

can both be cross-multiplied and solved to obtain $a = 5.6$ and $b = 4.8$.

A real-life situation that uses similar triangles involves measuring shadows to find heights of unknown objects. Consider the following problem: A building casts a shadow that is 120 feet long, and at the same time, another building that is 80 feet high casts a shadow that is 60 feet long. How tall is the first building? Each building, together with the sun rays and shadows casted on the ground, forms a triangle. They are similar because each building forms a right angle with the ground, and the sun rays form equivalent angles. Therefore, these two pairs of angles are both equal. Because all angles in a triangle add up to 180 degrees, the third angles are equal as well. Both shadows form corresponding sides of the triangle, the buildings form corresponding sides, and the sun rays form corresponding sides. Therefore, the triangles are similar, and the following proportion can be used to find the missing building length:

$$\frac{120}{x} = \frac{60}{80}$$

Cross-multiply to obtain the cross products, $9600 = 60x$. Then, divide both sides by 60 to obtain $x = 160$. This solution means that the other building is 160 feet high.

Solving Real-World Unit Rate Problems

A **unit rate** is a rate with a denominator of one. It is a comparison of two values with different units where one value is equal to one. Examples of unit rates include 60 miles per hour and 200 words per minute. Problems involving unit rates may require some work to find the unit rate. For example, if Mary travels 360 miles in 5 hours, what is her speed, expressed as a unit rate? The rate can be expressed as the following fraction:

$$\frac{360\ miles}{5\ hours}$$

The denominator can be changed to one by dividing by five. The numerator will also need to be divided by five to follow the rules of equality. This division turns the fraction into

$$\frac{72\ miles}{1\ hour}$$

which can now be labeled as a unit rate because one unit has a value of one. Another type question involves the use of unit rates to solve problems. For example, if Trey needs to read 300 pages and his average speed is 75 pages per hour, will he be able to finish the reading in 5 hours? The unit rate is 75 pages per hour, so the total of 300 pages can be divided by 75 to find the time. After the division, the time it takes to read is 4 hours. The answer to the question is yes, Trey will finish the reading within 5 hours.

Measures of Central Tendency

The three most common calculations for a set of data are the mean, median, and mode. These three are called **measures of central tendency**. Measures of central tendency are helpful in comparing two or more different sets of data. The most common measure of center is the **mean**, also referred to as the average.

To calculate the mean:

- Add all data values together.
- Divide by the sample size (the number of data points in the set).

The **median** is middle data value, so that half of the data lies below this value and half lies above it.

To calculate the median:

- Order the data from least to greatest.
- The point in the middle of the set is the median.
- If there is an even number of data points, add the two middle points and divide by 2.

The **mode** is the data value that occurs most often.

To calculate the mode:

- Order the data from least to greatest.
- Find the value that occurs most often.

Example: Amelia is a leading scorer on the school's basketball team. The following data set represents the number of points that Amelia has scored in each game this season. Use the mean, median, and mode to describe the data.

16, 12, 26, 14, 28, 14, 12, 15, 25

Solution:

Mean:

$$16 + 12 + 26 + 14 + 28 + 14 + 12 + 15 + 25 = 162$$

$$162 \div 9 = 18$$

Amelia averages 18 points per game.

Median:

12, 12, 14, 14, **15**, 16, 25, 26, 28

Amelia's median score is 15.

Mode:

12, 12, 14, 14, 15, 16, 25, 26, 28

The numbers 12 and 14 each occur twice, so this data set has 2 modes: 12 and 14.

The **range** is the difference between the largest and smallest values in the set. In the example above, the range is $28 - 12 = 16$.

An **outlier** is a data point that lies an unusual distance from other points in the data set. Removing an outlier from a data set will change the measures of center. Removing a **large outlier** (a high number) from a data set will decrease both the mean and the median. Removing a **small outlier** (a number much lower than most in the data set) from a data set will increase both the mean and the median. For example, given the data set {3, 6, 8, 12, 13, 14, 60}, the data point, 60, is an outlier because it is unusually far from the other points. In this data set, the mean is 16.6. Notice that this mean number is even larger than all other data points in the set except for 60. Removing the outlier, the mean changes to 9.3 and the median becomes 10. Removing an outlier will also decrease the range. In the data set above, the range is 57 when the outlier is included, but decreases to 11 when the outlier is removed.

Adding an outlier to a data set will affect the centers of measure as well. When a larger outlier is added to a data set, the mean and median increase. When a small outlier is added to a data set, the mean and median decrease. Adding an outlier to a data set will increase the range.

This does not seem to provide an appropriate measure of center when considering this data set. What will happen if that outlier is removed? Removing the extremely large data point, 60, is going to reduce the mean to 9.3. The mean decreased dramatically because 60 was much larger than any of the other data values. What would happen with an extremely low value in a data set like this one, {12, 87, 90, 95, 98, 100}? The mean of the given set is 80. When the outlier, 12, is removed, the mean should increase and should fit more closely to the other data points. Removing 12 and recalculating the mean shows that this is correct. The mean after removing 12 is 94. So, removing a large outlier will decrease the mean while removing a small outlier will increase the mean.

Modeling Numbers and Their Operations in Different Ways

Representing Rational Numbers on a Number Line

A **number line** is a tool used to compare numbers by showing where they fall in relation to one another. Labeling a number line with integers is simple because they have no fractional component and the values are easier to understand. The number line may start at -3 and go up to -2, then -1, then 0, and 1, 2, 3. This order shows that number 2 is larger than -1 because it falls further to the right on the number line. When positioning rational numbers, the process may take more time because it requires that they all be in the same form. If they are percentages, fractions, and decimals, then conversions will have to be made to put them in the same form. For example, if the numbers $\frac{5}{4}$, 45%, and 2.38 need to be put in order on a number line, the numbers must first be transformed into one single form.

Decimal form is an easy common ground because fractions can be changed by simply dividing, and percentages can be changed by moving the decimal point. After conversions are made, the list becomes 1.25, 0.45, and 2.38 respectively. Now the list is easier to arrange. The number line with the list in order is shown in the top half of the graphic below in the order 0.45, 1.25, and 2.38:

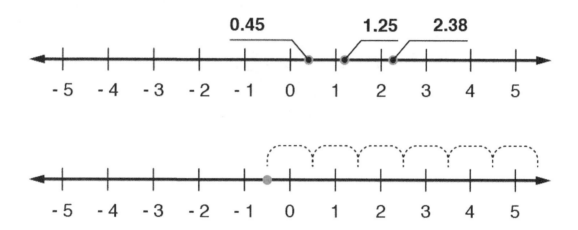

The sums and differences of rational numbers can be found using a number line after the rational numbers are put into the same form. This method is especially helpful when understanding the addition and subtraction of negative numbers. For example, the rational number six can be added to negative one-half using the number line. The following expression represents the problem:

$$-\frac{1}{2} + 6$$

First, the original number $-\frac{1}{2}$ can plotted by a dot on the number line, as shown in the lower half of the graphic above. Then 6 can be added by counting by whole numbers on the number line. The arcs on the graph represent the addition. The final answer is positive $5\frac{1}{2}$.

Producing, Interpreting, Understanding, Evaluating, and Improving Models

Concrete **models** create a way of thinking about math that generates learning on a more permanent level. Memorizing abstract math formulas will not create a lasting effect on the brain. The following picture shows fractions represented by Lego™ blocks. By starting with the whole block of eight, it can be split into half, which is a four-block, and a fourth, which is a two-block. The one-eighth representation is a single block.

After splitting these up, addition and subtraction can be performed by adding or taking away parts of the blocks. Different combinations of fractions can be used to make a whole, or taken away to make various parts of a whole.

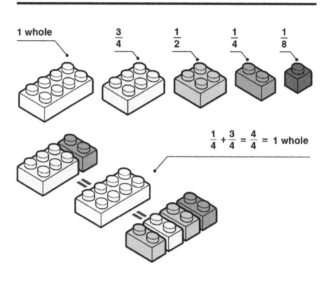

Using Colored Blocks to Model Functions

1 whole $\frac{3}{4}$ $\frac{1}{2}$ $\frac{1}{4}$ $\frac{1}{8}$

$\frac{1}{4} + \frac{3}{4} = \frac{4}{4} = 1$ whole

Multiplication can also be done using array models. The following picture shows a model of multiplying 3 times 4. **Arrays** are formed when the first factor is shown in a row. The second factor is shown in that number of columns. When the rectangle is formed, the blocks fill in to make a total, or the result of multiplication. The 3 rows and 4 columns show each factor and when the blocks are filled in, the total is 12. Arrays are great ways to represent multiplication because they show each factor, and where the total comes from, with rows and columns.

Multiplication Array Model for 3 x 4

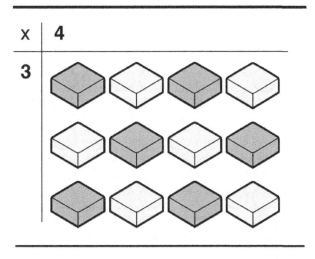

3 x 4 = 12

Another representation of fractions is shown below in the pie charts. Moving from whole numbers to part of numbers with fractions can be a concept that is difficult to grasp. Starting with a whole pie and splitting it into parts can be helpful with generating fractions. The first pie shows quarters or sections that are one-fourth because it is split into four parts. The second pie shows parts that equal one-fifth because it is split into five parts. Pies can also be used to add fractions. If $\frac{1}{5}$ and $\frac{1}{4}$ are being added, a common denominator must be found by splitting the pies into the same number of parts. The same number of parts can be found by determining the least common multiple. For 4 and 5, the least common multiple is 20. The pies can be split until there are 20 parts. The same portion of the pie can be shaded for each fraction and then added together to find the sum.

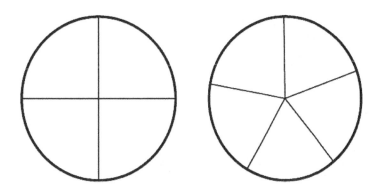

Multiplication and Division Problems Using Equations, Rectangular Arrays, and Area Models

Multiplication and division can be represented by equations. These equations show the numbers involved and the operation. For example, "eight multiplied by six equals forty-eight" is seen in the following equation:

$$8 \times 6 = 48$$

This operation can be modeled by rectangular arrays where one factor, 8, is the number of rows, and the other factor, 6, is the number of columns, as follows:

Array of 8 x 6 = 48

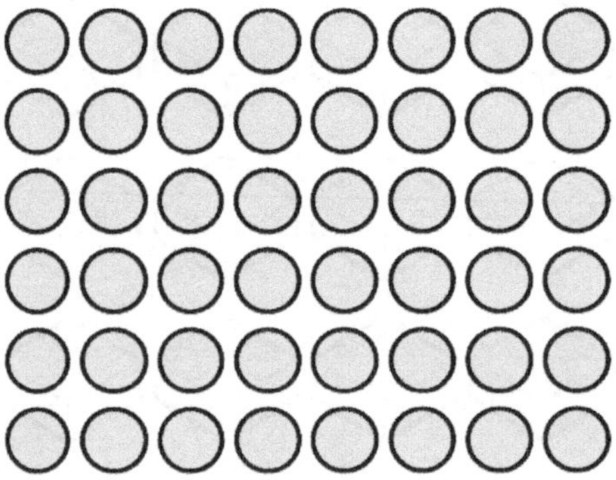

Rectangular arrays show what happens with the concept of multiplication. The first factor in the problem is used to draw the first row of dots. Then the second factor is used to add the number of columns. The final model includes 6 rows of 8 columns, which results in 48 dots. These rectangular arrays show how multiplication of whole numbers will result in a number larger than the factors.

Division can also be represented by equations and area models. A division problem such as "twenty-four divided by three equals eight" can be written as the following equation: $24 \div 8 = 3$. The object below is an area model used to represent the equation. As seen in the model, the whole box represents 24 and the 3 sections represent the division by 3. In more detail, there could be 24 dots written in the whole box, and each box could have 8 dots in it. Division shows how numbers can be divided into groups. For the example problem, it is asking how many numbers will be in each of the 3 groups that together make 24. The answer is 8 in each group.

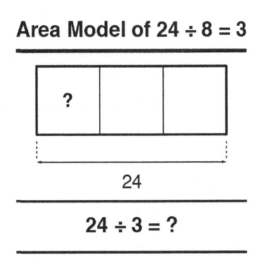

Area Model of 24 ÷ 8 = 3

$$24 \div 3 = ?$$

Number Manipulation

Real and Complex Number Systems

Whole numbers are the numbers 0, 1, 2, 3, …. Examples of other whole numbers would be 413 and 8,431. Notice that numbers such as 4.13 and $\frac{1}{4}$ are not included in whole numbers. **Counting numbers**, also known as **natural numbers**, consist of all whole numbers except for the zero. In set notation, the natural numbers are the set $\{1, 2, 3, …\}$. The entire set of whole numbers and negative versions of those same numbers comprise the set of numbers known as **integers**. Therefore, in set notation, the integers are $\{…, -3, -2, -1, 0, 1, 2, 3, …\}$. Examples of other integers are $-4{,}981$ and $90{,}131$. A number line is a great way to visualize the integers. Integers are labeled on the following number line:

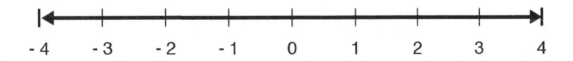

The arrows on the right- and left-hand sides of the number line show that the line continues indefinitely in both directions.

Fractions also exist on the number line as parts of a whole. For example, if an entire pie is cut into two pieces, each piece is half of the pie, or $\frac{1}{2}$. The top number in any fraction, known as the **numerator**, defines how many parts there are. The bottom number, known as the **denominator**, states how many pieces the whole is divided into. Fractions can also be negative or written in their corresponding decimal form.

A **decimal** is a number that uses a decimal point and numbers to the right of the decimal point representing the part of the number that is less than 1. For example, 3.5 is a decimal and is equivalent to the fraction $\frac{7}{2}$ or the mixed number $3\frac{1}{2}$. The decimal is found by dividing 2 into 7. Other examples of fractions are $\frac{2}{7}$, $\frac{-3}{14}$, and $\frac{14}{27}$.

Any number that can be expressed as a fraction is known as a **rational number**. Basically, if a and b are any integers and $b \neq 0$, then $\frac{a}{b}$ is a rational number. Any integer can be written as a fraction where the denominator is 1, so therefore the rational numbers consist of all fractions and all integers.

Any number that is not rational is known as an **irrational number**. Consider the number:

$$\pi = 3.141592654\ldots.$$

The decimal portion of that number extends indefinitely. In that situation, a number can never be written as a fraction. Another example of an irrational number is:

$$\sqrt{2} = 1.414213662\ldots.$$

Again, this number cannot be written as a ratio of two integers.

Together, the set of all rational and irrational numbers makes up the real numbers**.** The number line contains all real numbers. To graph a number other than an integer on a number line, it needs to be plotted between two integers. For example, 3.5 would be plotted halfway between 3 and 4.

Even numbers are integers that are divisible by 2. For example, 6, 100, 0, and −200 are all even numbers. **Odd numbers** are integers that are not divisible by 2. If an odd number is divided by 2, the result is a fraction. For example, −5, 11, and −121 are odd numbers.

Prime numbers consist of natural numbers greater than 1 that are not divisible by any other natural numbers other than themselves and 1. For example, 3, 5, and 7 are prime numbers. If a natural number is not prime, it is known as a **composite number**. Eight is a composite number because it is divisible by both 2 and 4, which are natural numbers other than itself and 1.

The **absolute value** of any real number is the distance from that number to zero on the number line. The absolute value of a number can never be negative. For example, the absolute value of both 8 and −8 is 8 because they are both 8 units away from zero on the number line. This is written as $|8| = |-8| = 8$.

Base-10 Numerals, Number Names, and Expanded Form

The **base-10 number system** is also called the **decimal system of naming numbers**. There is a decimal point that sets the value of numbers based on their position relative to the decimal point. The order from the decimal point to the right is the tenths place, then hundredths place, then thousandths place. Moving to the left from the decimal point, the place value is ones, tens, hundreds, etc. The number 2,356 can be described in words as "two thousand three hundred fifty-six." In expanded form, it can be written as:

$$(2 \times 1{,}000) + (3 \times 100) + (5 \times 10) + (6 \times 1)$$

The expanded form shows the value each number holds in its place. The number 3,093 can be written in words as "three thousand ninety-three." In expanded form, it can be expressed as:

$$(3 \times 1,000) + (0 \times 100) + (9 \times 10) + (3 \times 1)$$

Notice that the zero is added in the expanded form as a place holder. There are no hundreds in the number, so a zero is written in the hundreds place.

Composing and Decomposing Multidigit Numbers

Composing and decomposing numbers reveals the place value held by each number zero through 9 in each position. For example, the number 17 is read as "seventeen." It can be decomposed into the numbers 10 and 7. It can be described as 1 group of ten and 7 ones. The one in the tens place represents one set of ten. The seven in the ones place represents seven sets of one. Added together, they make a total of seventeen. The number 48 can be written in words as "forty-eight." It can be decomposed into the numbers 40 and 8, where there are 4 groups of ten and 8 groups of one. The number 296 can be decomposed into 2 groups of one hundred, 9 groups of ten, and 6 groups of one. There are two hundreds, nine tens, and six ones. Decomposing and composing numbers lays the foundation for visually picturing the number and its place value, and adding and subtracting multiple numbers with ease.

Base-10 System

The Place and Value of a Digit

Each number in the base-10 system is made of the numbers zero through 9, located in different places relative to the decimal point. Based on where the numbers fall, the value of a digit changes. For example, the number 7,509 has a seven in the thousands place. This means there are seven groups of one thousand. The number 457 has a seven in the ones place. This means there are seven groups of one. Even though there is a seven in both numbers, the place of the seven tells the value of the digit. A practice question may ask the place and value of the 4 in 3,948. The four is found in the tens place, which means four represents the number 40, or four groups of ten. Another place value may be on the opposite side of the decimal point. A question may ask the place and value of the 8 in the number 203.80. In this case, the eight is in the tenths place because it is in the first place to the right of the decimal point. It holds a value of eight-tenths, or eight groups of one-tenth.

The Relative Value of a Digit

The value of a digit is found by recognizing its place relative to the rest of the number. For example, the number 569.23 contains a 6. The position of the 6 is two places to the left of the decimal, putting it in the tens place. The tens place gives it a value of 60, or six groups of ten. The number 39.674 has a 4 in it. The number 4 is located three places to the right of the decimal point, placing it in the thousandths place. The value of the 4 is four-thousandths, because of its position relative to the other numbers and to the decimal. It can be described as 0.004 by itself, or four groups of one-thousandths. The numbers 100 and 0.1 are both made up of ones and zeros. The first number, 100, has a 1 in the hundreds place, giving it a value of one hundred. The second number, 0.1, has a 1 in the tenths place, giving that 1 a value of one-tenth. The place of the number gives it the value.

Rounding Multidigit Numbers

Numbers can be rounded by recognizing the place value where the rounding takes place, then looking at the number to the right. If the number to the right is 5 or greater, the number to be rounded goes up one. If the number to the right is 4 or less, the number to be rounded stays the same. For example, the number 438 can be rounded to the tens place. The number 3 is in the tens place and the number to the right is 8. Because the 8 is 5 or greater, the 3 then rounds up to a 4. The rounded number is 440. Another number, 1,394, can be rounded to the thousands place. The number in the thousands place is 1, and the number to the right is 3. As the 3 is 4 or less, it means the 1 stays the same, and the rounded number is 1,000. Rounding is also a form of estimating. The number 9.58 can be rounded to the tenths place. The number 5 is in the tenths place, and the number 8 is to the right of it. Because 8 is 5 or greater, the 5 changes to a 6. The rounded number becomes 9.6.

Rational Exponents

Rational exponents are used to express the root of a number raised to a specific power. For example, $3^{\frac{1}{2}}$ has a base of 3 and rational exponent of $\frac{1}{2}$. The square root of 3 raised to the first power can be written as $\sqrt[2]{3^1}$. Any number with a rational exponent can be written this way.

The **numerator**, or number on top of the fraction, becomes the whole number exponent and the **denominator**, or bottom number of the fraction, becomes the root. Another example is $4^{\frac{3}{2}}$. It can be rewritten as the square root of four to the third power, or $\sqrt[2]{4^3}$. This can be simplified by performing the operations 4 to the third power:

$$4^3 = 4 \times 4 \times 4 = 64$$

and then taking the square root of 64, $\sqrt[2]{64}$, which yields an answer of 8. Another way of stating the answer would be 4 to power of $\frac{3}{2}$ is eight, or that 4 to the power of $\frac{3}{2}$ is the square root of 4 cubed:

$$\sqrt[2]{4}^3 = 2^3 = 2 \times 2 \times 2 = 8$$

Basic Concepts of Number Theory

Prime and Composite Numbers

A prime number is a whole number greater than 1 that can only be divided by 1 and itself. Examples are 2, 3, 5, 7, and 11. A composite number can be evenly divided by a number other than 1 and itself. Examples of composite numbers are 4 and 9. Four can be divided evenly by 1, 2, and 4. Nine can be divided evenly by 1, 3, and 9. When given a list of numbers, one way to determine which ones are prime or composite is to find the prime factorization of each number. For example, a list of numbers may include 13 and 24. The prime factorization of 13 is 1 and 13 because those are the only numbers that go into it evenly, so it is a prime number. The prime factorization of 24 is:

$$2 \times 2 \times 2 \times 3$$

because those are the prime numbers that multiply together to get 24. This also shows that 24 is a composite number because 2 and 3 are factors along with 1 and 24.

Multiples

A **multiple** of a number is the result of multiplying that number by an integer. For example, some multiples of 3 are 6, 9, 12, 15, and 18. These multiples are found by multiplying 3 by 2, 3, 4, 5, and 6, respectively. Some multiples of 5 include 5, 10, 15, and 20. This also means that 5 is a factor of 5, 10, 15, and 20. Some questions may ask which numbers in a list are multiples of a given number. For example, find and circle the multiples of 12 in the following list: 136, 144, 312, 400. If a number is evenly divisible by 12, then it is a multiple of 12. The numbers 144 and 312 are multiples of 12 because 12 times 12 is 144, and 12 times 26 is 312. The other numbers, 136 and 400, are not multiples because they yield a number with a fractional component when divided by 12.

Factorization

Factorization is the process of breaking up a mathematical quantity, such as a number or polynomial, into a product of two or more factors. For example, a factorization of the number 16 is:

$$16 = 8 \times 2$$

If multiplied out, the factorization results in the original number. A **prime factorization** is a specific factorization when the number is factored completely using prime numbers only. For example, the prime factorization of 16 is:

$$16 = 2 \times 2 \times 2 \times 2$$

A factor tree can be used to find the prime factorization of any number. Within a factor tree, pairs of factors are found until no other factors can be used, as in the following factor tree of the number 84:

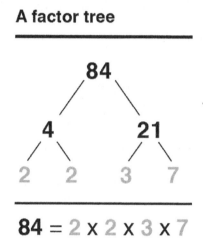

A factor tree

$$84 = 2 \times 2 \times 3 \times 7$$

It first breaks 84 into 21 × 4, which is not a prime factorization. Then, both 21 and 4 are factored into their primes. The final numbers on each branch consist of the numbers within the prime factorization. Therefore:

$$84 = 2 \times 2 \times 3 \times 7$$

Factorization can be helpful in finding greatest common divisors and least common denominators.

Also, a factorization of an algebraic expression can be found. Throughout the process, a more complicated expression can be decomposed into products of simpler expressions. To factor a polynomial, first determine if there is a greatest common factor. If there is, factor it out. For example:

$$2x^2 + 8x$$

has a greatest common factor of $2x$ and can be written as:

$$2x(x + 4)$$

Once the greatest common monomial factor is factored out, if applicable, count the number of terms in the polynomial. If there are two terms, is it a difference of squares, a sum of cubes, or a difference of cubes?

If so, the following rules can be used:

Difference of Squares:

$$a^2 - b^2 = (a + b)(a - b)$$

Sum of Cubes:

$$a^3 + b^3 = (a + b)(a^2 - ab + b^2)$$

Difference of Cubes

$$a^3 - b^3 = (a - b)(a^2 + ab + b^2)$$

If there are three terms, and if the trinomial is a perfect square trinomial, it can be factored into the following:

$$a^2 + 2ab + b^2 = (a + b)^2$$

$$a^2 - 2ab + b^2 = (a - b)^2$$

If not, try factoring into a product of two binomials by trial and error into a form of $(x + p)(x + q)$. For example, to factor:

$$x^2 + 6x + 8$$

determine what two numbers have a product of 8 and a sum of 6. Those numbers are 4 and 2, so the trinomial factors into:

$$(x + 2)(x + 4)$$

Finally, if there are four terms, try factoring by grouping. First, group terms together that have a common monomial factor. Then, factor out the common monomial factor from the first two terms. Next, look to see if a common factor can be factored out of the second set of two terms that results in a common binomial factor. Finally, factor out the common binomial factor of each expression, for example:

$$xy - x + 5y - 5$$

$$x(y - 1) + 5(y - 1) = (y - 1)(x + 5)$$

After the expression is completely factored, check to see if the factorization is correct by multiplying to try to obtain the original expression. Factorizations are helpful in solving equations that consist of a polynomial set equal to zero. If the product of two algebraic expressions equals zero, then at least one of the factors is equal to zero. Therefore, factor the polynomial within the equation, set each factor equal to zero, and solve.

For example:

$$x^2 + 7x - 18 = 0$$

can be solved by factoring into:

$$(x + 9)(x - 2) = 0$$

Set each factor equal to zero, and solve to obtain $x = -9$ and $x = 2$.

Ordering and Comparing Rational Numbers
Ordering rational numbers is a way to compare two or more different numerical values. Determining whether two amounts are equal, less than, or greater than is the basis for comparing both positive and negative numbers. Also, a group of numbers can be compared by ordering them from the smallest amount to the largest amount. A few symbols are necessary to use when ordering rational numbers. The equals sign, =, shows that the two quantities on either side of the symbol have the same value. For example:

$$\frac{12}{3} = 4$$

because both values are equivalent. Another symbol that is used to compare numbers is $<$, which represents "less than." With this symbol, the smaller number is placed on the left and the larger number is placed on the right. Always remember that the symbol's "mouth" opens up to the larger number. When comparing negative and positive numbers, it is important to remember that the number occurring to the left on the number line is always smaller and is placed to the left of the symbol. This idea might seem confusing because some values could appear at first glance to be larger, even though they are not. For example, $-5 < 4$ is read "negative 5 is less than 4." Here is an image of a number line for help:

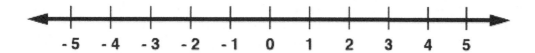

The symbol $\leq$ represents "less than or equal to," and it joins $<$ with equality. Therefore, both:

$$-5 \leq 4$$

and

$$-5 \leq -5$$

are true statements and "-5 is less than or equal to both 4 and -5." Other symbols are $>$ and $\geq$, which represent "greater than" and "greater than or equal to." Both:

$$4 \geq -1$$

and

$$-1 \geq -1$$

are correct ways to use these symbols.

Here is a chart of these four inequality symbols:

Symbol	Definition
<	less than
≤	less than or equal to
>	greater than
≥	greater than or equal to

Comparing integers is a straightforward process, especially when using the number line, but the comparison of decimals and fractions is not as obvious. When comparing two non-negative decimals, compare digit by digit, starting from the left. The larger value contains the first larger digit. For example, 0.1456 is larger than 0.1234 because the value 4 in the hundredths place in the first decimal is larger than the value 2 in the hundredths place in the second decimal. When comparing a fraction with a decimal, convert the fraction to a decimal and then compare in the same manner. Finally, there are a few options when comparing fractions. If two non-negative fractions have the same denominator, the fraction with the larger numerator is the larger value.

If they have different denominators, they can be converted to equivalent fractions with a common denominator to be compared, or they can be converted to decimals to be compared. When comparing two negative decimals or fractions, a different approach must be used. It is important to remember that the smaller number exists to the left on the number line. Therefore, when comparing two negative decimals by place value, the number with the larger first place value is smaller due to the negative sign. Whichever value is closer to zero is larger. For instance, -0.456 is larger than -0.498 because of the values in the hundredth places. If two negative fractions have the same denominator, the fraction with the larger numerator is smaller because of the negative sign.

Converting Non-Negative Fractions, Decimals, and Percentages

Within the number system, different forms of numbers can be used. It is important to be able to recognize each type, as well as work with, and convert between, the given forms. The **real number system** comprises natural numbers, whole numbers, integers, rational numbers, and irrational numbers. Natural numbers, whole numbers, integers, and irrational numbers typically are not represented as fractions, decimals, or percentages. Rational numbers, however, can be represented as any of these three forms. A rational number is a number that can be written in the form $\frac{a}{b}$, where a and b are integers, and b is not equal to zero. In other words, rational numbers can be written in a fraction form. The value a is the numerator, and b is the denominator. If the numerator is equal to zero, the entire fraction is equal to zero. Non-negative fractions can be less than 1, equal to 1, or greater than 1. Fractions are less than 1 if the numerator is smaller (less than) than the denominator.

For example, $\frac{3}{4}$ is less than 1. A fraction is equal to 1 if the numerator is equal to the denominator. For instance, $\frac{4}{4}$ is equal to 1. Finally, a fraction is greater than 1 if the numerator is greater than the denominator: the fraction $\frac{11}{4}$ is greater than 1. When the numerator is greater than the denominator, the fraction is called an **improper fraction**. An improper fraction can be converted to a **mixed number**, a combination of both a whole number and a fraction. To convert an improper fraction to a mixed number, divide the numerator by the denominator. Write down the whole number portion, and then write any remainder over the original denominator. For example, $\frac{11}{4}$ is equivalent to $2\frac{3}{4}$. Conversely, a mixed number can be converted to an improper fraction by multiplying the denominator by the whole number and adding that result to the numerator.

Fractions can be converted to decimals. With a calculator, a fraction is converted to a decimal by dividing the numerator by the denominator. For example:

$$\frac{2}{5} = 2 \div 5 = 0.4$$

Sometimes, rounding might be necessary. Consider:

$$\frac{2}{7} = 2 \div 7 = 0.28571429$$

This decimal could be rounded for ease of use, and if it needed to be rounded to the nearest thousandth, the result would be 0.286. If a calculator is not available, a fraction can be converted to a decimal manually. First, find a number that, when multiplied by the denominator, has a value equal to 10, 100, 1,000, etc. Then, multiply both the numerator and denominator times that number. The decimal form of the fraction is equal to the new numerator with a decimal point placed as many place values to the left as there are zeros in the denominator. For example, to convert $\frac{3}{5}$ to a decimal, multiply both the numerator and denominator times 2, which results in $\frac{6}{10}$.

The decimal is equal to 0.6 because there is one zero in the denominator, and so the decimal place in the numerator is moved one unit to the left. In the case where rounding would be necessary while working without a calculator, an approximation must be found. A number close to 10, 100, 1,000, etc. can be used. For example, to convert $\frac{1}{3}$ to a decimal, the numerator and denominator can be multiplied by 33 to turn the denominator into approximately 100, which makes for an easier conversion to the equivalent decimal. This process results in $\frac{33}{99}$ and an approximate decimal of 0.33. Once in decimal form, the number can be converted to a percentage. Multiply the decimal by 100 and then place a percent sign after the number. For example, 0.614 is equal to 61.4%. In other words, move the decimal place two units to the right and add the percent symbol.

Arithmetic Operations with Rational Numbers

Addition, Subtraction, Multiplication, and Division with Whole Numbers
The four basic operations include addition, subtraction, multiplication, and division. The result of addition is a **sum**, the result of subtraction is a **difference**, the result of multiplication is a **product**, and the result of division is a **quotient**. Each type of operation can be used when working with rational numbers; however, the basic operations need to be understood first while using simpler numbers before working with fractions and decimals.

Performing these operations should first be learned using whole numbers. Addition needs to be done column by column. To add two whole numbers, add the ones column first, then the tens columns, then the hundreds, etc. If the sum of any column is greater than 9, a 1 must be carried over to the next column. For example, the following is the result of $482 + 924$:

$$\begin{array}{r} {\scriptstyle 1} \\ 482 \\ +924 \\ \hline 1406 \end{array}$$

Notice that the sum of the tens column was 10, so a one was carried over to the hundreds column. Subtraction is also performed column by column. Subtraction is performed in the ones column first, then the tens, etc. If the number on top is less than the number below, a one must be borrowed from the column to the left. For example, the following is the result of $5,424 - 756$:

$$\begin{array}{r} 4 \ 13 \ 11 \ 14 \\ \cancel{5} \ \cancel{4} \ \cancel{2} \ 4 \\ - \ 7 \ 5 \ 6 \\ \hline 4 \ 6 \ 6 \ 8 \end{array}$$

Notice that a one is borrowed from the tens, hundreds, and thousands place. After subtraction, the answer can be checked through addition. A check of this problem would be to show that $756 + 4,668 = 5,424$.

Multiplication of two whole numbers is performed by writing one on top of the other. The number on top is known as the **multiplicand**, and the number below is the **multiplier**. Perform the multiplication by multiplying the multiplicand by each digit of the multiplier. Make sure to place the ones value of each result under the multiplying digit in the multiplier. Each value to the right is then a zero. The product is found by adding each product. For example, the following is the process of multiplying 46 times 37, where 46 is the multiplicand and 37 is the multiplier:

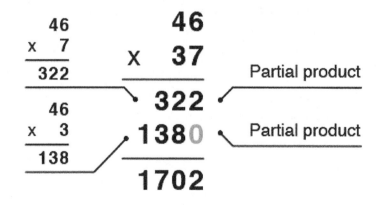

Finally, division can be performed using long division. When dividing a number by another number, the first number is known as the **dividend**, and the second is the **divisor**. For example, with $a \div b = c$, a is the dividend, b is the divisor, and c is the quotient. For long division, place the dividend within the division symbol and the divisor on the outside. For example, with $8{,}764 \div 4$, refer to the first problem in the diagram below. First, there are two 4's in the first digit, 8. This number 2 gets written above the 8. Then, multiply 4 times 2 to get 8, and that product goes below the 8. Subtract to get 8, and then carry down the 7. Continue the same steps.

$$7 \div 4 = 1 \text{ R3}$$

so 1 is written above the 7. Multiply 4 times 1 to get 4, and write it below the 7. Subtract to get 3, and carry the 6 down next to the 3. Resulting steps give a 9 and a 1. The final subtraction results in a zero, which means that 8,764 is divisible by 4. There are no remaining numbers.

The second example shows that:

$$4{,}536 \div 216 = 21$$

The steps are a little different because 216 cannot be contained in 4 or 5, so the first step is placing a 2 above the 3 because there are two 216's in 453. Finally, the third example shows that:

$$546 \div 31 = 17 \text{ R19}$$

The 19 is a remainder. Notice that the final subtraction does not result in a zero, which means that 546 is not divisible by 31. The remainder can also be written as a fraction over the divisor to say that:

$$546 \div 31 = 17\frac{19}{31}$$

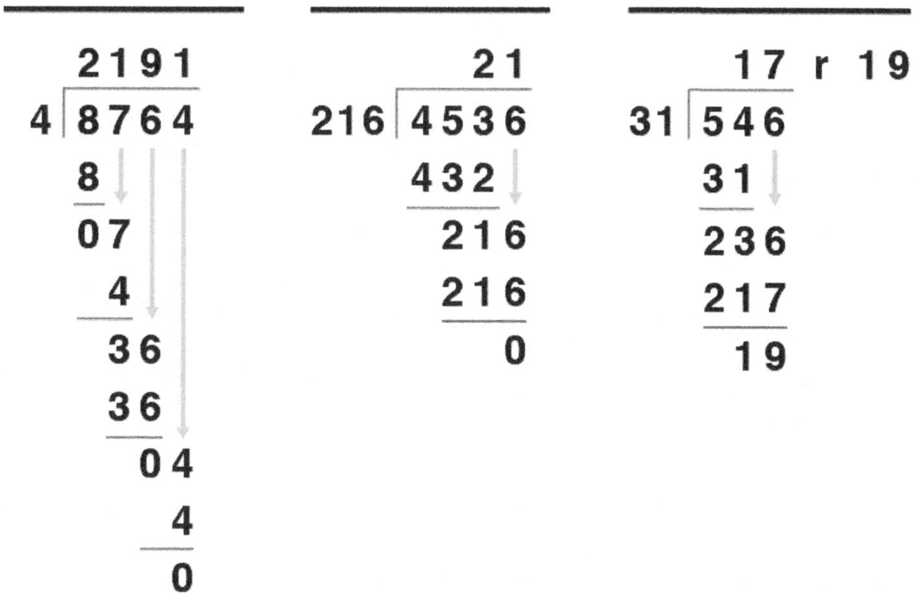

$$
\begin{array}{r}
2\,1\,9\,1 \\
4\,\overline{)\,8\,7\,6\,4} \\
8 \\
\overline{0\,7} \\
4 \\
\overline{3\,6} \\
3\,6 \\
\overline{0\,4} \\
4 \\
\overline{0}
\end{array}
\qquad
\begin{array}{r}
2\,1 \\
216\,\overline{)\,4\,5\,3\,6} \\
4\,3\,2 \\
\overline{2\,1\,6} \\
2\,1\,6 \\
\overline{0}
\end{array}
\qquad
\begin{array}{r}
1\,7 \ r \ 1\,9 \\
31\,\overline{)\,5\,4\,6} \\
3\,1 \\
\overline{2\,3\,6} \\
2\,1\,7 \\
\overline{1\,9}
\end{array}
$$

If a division problem relates to a real-world application, and a remainder does exist, it can have meaning. For example, consider the third example:

$$546 \div 31 = 17 \text{ R19}$$

Let's say that we had \$546 to spend on calculators that cost \$31 each, and we wanted to know how many we could buy. The division problem would answer this question. The result states that 17 calculators could be purchased, with \$19 left over. Notice that the remainder will never be greater than or equal to the divisor.

Addition, Subtraction, Multiplication, and Division with Positive and Negative Numbers

Once the operations are understood with whole numbers, they can be used with integers. There are many rules surrounding operations with negative numbers. First, consider addition with integers. The sum of two numbers can first be shown using a number line. For example, to add:

$$-5 + (-6)$$

plot the point −5 on the number line. Then, because a negative number is being added, move 6 units to the left. This process results in landing on −11 on the number line, which is the sum of −5 and −6. If adding a positive number, move to the right.

Visualizing this process using a number line is useful for understanding; however, it is not efficient. A quicker process is to learn the rules. When adding two numbers with the same sign, add the absolute values of both numbers, and use the common sign of both numbers as the sign of the sum. For example, to add:

$$-5 + (-6)$$

add their absolute values:

$$5 + 6 = 11$$

Then, introduce a negative number because both addends are negative. The result is −11. To add two integers with unlike signs, subtract the lesser absolute value from the greater absolute value, and apply the sign of the number with the greater absolute value to the result. For example, the sum $-7 + 4$ can be computed by finding the difference:

$$7 - 4 = 3$$

and then applying a negative because the value with the larger absolute value is negative. The result is −3. Similarly, the sum $-4 + 7$ can be found by computing the same difference but leaving it as a positive result because the addend with the larger absolute value is positive. Also, recall that any number plus zero equals that number. This is known as the **Addition Property of Zero**.

Subtracting two integers can be computed by changing to addition to avoid confusion. The rule is to add the first number to the opposite of the second number. The opposite of a number is the number on the other side of zero on the number line, which is the same number of units away from zero. For example, −2 and 2 are opposites. Consider $4 - 8$. Change this to adding the opposite as follows:

$$4 + (-8)$$

Then, follow the rules of addition of integers to obtain −4. Secondly, consider:

$$-8 - (-2)$$

Change this problem to adding the opposite as $-8 + 2$, which equals −6. Notice that subtracting a negative number functions the same as adding a positive number.

Multiplication and division of integers are actually less confusing than addition and subtraction because the rules are simpler to understand. If two factors in a multiplication problem have the same sign, the result is positive. If one factor is positive and one factor is negative, the result, known as the product, is negative. For example:

$$(-9)(-3) = 27$$

and

$$9(-3) = -27$$

Also, a number times zero always results in zero. If a problem consists of more than a single multiplication, the result is negative if it contains an odd number of negative factors, and the result is positive if it contains an even number of negative factors. For example:

$$(-1)(-1)(-1)(-1) = 1$$

and

$$(-1)(-1)(-1)(-1)(-1) = -1$$

These two examples of multiplication also bring up another concept. Both are examples of repeated multiplication, which can be written in a more compact notation using exponents. The first example can be written as:

$$(-1)^4 = 1$$

and the second example can be written as:

$$(-1)^5 = -1$$

Both are exponential expressions: −1 is the base in both instances, and 4 and 5 are the respective exponents. Note that a negative number raised to an odd power is always negative, and a negative number raised to an even power is always positive. Also, $(-1)^4$ is not the same as -1^4. In the first expression, the negative is included in the parentheses, but it is not in the second expression. The second expression is found by evaluating 1^4 first to get 1 and then applying the negative sign to obtain −1.

A similar theory applies within division. First, consider some vocabulary. When dividing 14 by 2, it can be written in the following ways:

$$14 \div 2 = 7$$

or

$$\frac{14}{2} = 7$$

Fourteen is the dividend, 2 is the divisor, and 7 is the quotient. If two numbers in a division problem have the same sign, the quotient is positive. If two numbers in a division problem have different signs, the quotient is negative. For example:

$$14 \div (-2) = -7, \text{ and } -14 \div (-2) = 7$$

To check division, multiply the quotient times the divisor to obtain the dividend. Also, remember that zero divided by any number is equal to zero. However, any number divided by zero is undefined. It does not make sense to divide a number by zero parts.

If more than one operation is to be completed in a problem, follow the **Order of Operations**. The mnemonic device, **PEMDAS,** for the order of operations states the order in which addition, subtraction, multiplication, and division need to be done. It also includes when to evaluate operations within grouping symbols and when to incorporate exponents. PEMDAS, which some remember by thinking "Please Excuse My Dear Aunt Sally," refers to parentheses, exponents, multiplication, division, addition, and subtraction. First, within an expression, complete any operation that is within parentheses, or any other grouping symbol like brackets, braces, or absolute value symbols. Note that this does not refer to the case when parentheses are used to represent multiplication like $(2)(5)$, wherein an operation is not within parentheses like it is in (2×5). Then, any exponents must be computed. Next, multiplication and division are performed from left to right.

Finally, addition and subtraction are performed from left to right. The following is an example in which the operations within the parentheses need to be performed first, so the order of operations must be applied to the exponent, subtraction, addition, and multiplication within the grouping symbol:

$$9 - 3(3^2 - 3 + 4 \cdot 3)$$

$$9 - 3(3^2 - 3 + 4 \cdot 3) \quad \text{Work within the parentheses first}$$

$$= 9 - 3(9 - 3 + 12)$$

$$= 9 - 3(18)$$

$$= 9 - 54$$

$$= -45$$

Addition, Subtraction, Multiplication, and Division with Fractions

Once the rules for integers are understood, operations with fractions and decimals can be mastered. Recall that a rational number can be written as a fraction and can be converted to a decimal through division. If a rational number is negative, the rules for adding, subtracting, multiplying, and dividing integers must be used. If a rational number is in fraction form, performing addition, subtraction, multiplication, and division is more complicated than when working with integers. First, consider addition. To add two fractions having the same denominator, add the numerators and then reduce the fraction.

When an answer is a fraction, it should always be in lowest terms. **Lowest terms** means that every common factor, other than 1, between the numerator and denominator is divided out. For example:

$$\frac{2}{8} + \frac{4}{8} = \frac{6}{8} = \frac{6 \div 2}{8 \div 2} = \frac{3}{4}$$

Both the numerator and denominator of $\frac{6}{8}$ have a common factor of 2, so 2 is divided out of each number to put the fraction in lowest terms. If denominators are different in an addition problem, the fractions must be converted to have common denominators. The **least common denominator (LCD)** of all the given denominators must be found, and this value is equal to the **least common multiple (LCM)** of the denominators. This non-zero value is the smallest number that is a multiple of both denominators. Then, each original fraction can be written as an equivalent fraction using the new denominator. Once in this form, process of adding with like denominators is completed. For example, consider:

$$\frac{1}{3} + \frac{4}{9}$$

The LCD is 9 because 9 is the smallest multiple of both 3 and 9. The fraction $\frac{1}{3}$ must be rewritten with 9 as its denominator. Therefore, multiply both the numerator and denominator by 3. Multiplying by $\frac{3}{3}$ is the same as multiplying times 1, which does not change the value of the fraction. Therefore, an equivalent fraction is $\frac{3}{9}$, and:

$$\frac{1}{3} + \frac{4}{9} = \frac{3}{9} + \frac{4}{9} = \frac{7}{9}$$

which is in lowest terms. Subtraction is performed in a similar manner; once the denominators are equal, the numerators are then subtracted. The following is an example of addition of a positive and a negative fraction:

$$-\frac{5}{12} + \frac{5}{9} = -\frac{5 \times 3}{12 \times 3} + \frac{5 \times 4}{9 \times 4}$$

$$-\frac{15}{36} + \frac{20}{36} = \frac{5}{36}$$

Common denominators are not used in multiplication and division. To multiply two fractions, multiply the numerators together and the denominators together. Then, write the result in lowest terms.

For example:

$$\frac{2}{3} \times \frac{9}{4} = \frac{18}{12} = \frac{3}{2}$$

Alternatively, the fractions could be factored first to cancel out any common factors before performing the multiplication. For example:

$$\frac{2}{3} \times \frac{9}{4} = \frac{2}{3} \times \frac{3 \times 3}{2 \times 2} = \frac{3}{2}$$

This second approach is helpful when working with larger numbers, as common factors might not be obvious. Multiplication and division of fractions are related because the division of two fractions is changed into a multiplication problem. This means that dividing a fraction by another fraction is the same as multiplying the first fraction by the reciprocal of the second fraction, so that second fraction must be inverted, or "flipped," to be in reciprocal form. For example:

$$\frac{11}{15} \div \frac{3}{5} = \frac{11}{15} \times \frac{5}{3} = \frac{55}{45} = \frac{11}{9}$$

The fraction $\frac{5}{3}$ is the reciprocal of $\frac{3}{5}$. It is possible to multiply and divide numbers containing a mix of integers and fractions. In this case, convert the integer to a fraction by placing it over a denominator of 1. For example, a division problem involving an integer and a fraction is:

$$3 \div \frac{1}{2} = \frac{3}{1} \times \frac{2}{1} = \frac{6}{1} = 6$$

Finally, when performing operations with rational numbers that are negative, the same rules apply as when performing operations with integers. For example, a negative fraction times a negative fraction results in a positive value, and a negative fraction subtracted from a negative fraction results in a negative value.

Addition, Subtraction, Multiplication, and Division with Decimals
Operations can be performed on rational numbers in decimal form. Recall that to write a fraction as an equivalent decimal expression, divide the numerator by the denominator. For example:

$$\frac{1}{8} = 1 \div 8 = 0.125$$

With the case of decimals, it is important to keep track of place value. To add decimals, make sure the decimal places are in alignment so that the numbers are lined up with their decimal points and add vertically. If the numbers do not line up because there are extra or missing place values in one of the numbers, then zeros may be used as placeholders. For example, $0.123 + 0.23$ becomes:

$$
\begin{array}{r}
0.123 \\
\underline{0.230} \\
0.353
\end{array}
$$

Subtraction is done the same way. Multiplication and division are more complicated. To multiply two decimals, place one on top of the other as in a regular multiplication process and do not worry about lining up the decimal points. Then, multiply as with whole numbers, ignoring the decimals. Finally, in the solution, insert the decimal point as many places to the left as there are total decimal values in the original problem. Here is an example of a decimal multiplication problem:

$$
\begin{array}{rl}
0.52 & \textit{2 decimal places} \\
\times\ \underline{0.2} & \textit{1 decimal place} \\
0.104 & \textit{3 decimal places}
\end{array}
$$

The answer to 52 times 2 is 104, and because there are three decimal values in the problem, the decimal point is positioned three units to the left in the answer.

The decimal point plays an integral role throughout the whole problem when dividing with decimals. First, set up the problem in a long division format. If the divisor is not an integer, the decimal must be moved to the right as many units as needed to make it an integer. The decimal in the dividend must be moved to the right the same number of places to maintain equality. Then, division is completed normally. Here is an example of long division with decimals:

$$
\begin{array}{r}
\text{Long division} \\
\text{with decimals} \\
\hline
0.06\,\overline{)\,12.72} \\
\end{array}
$$

$$
\begin{array}{r}
212 \\
6\,\overline{)\,1272} \\
12\downarrow \\
\hline
07 \\
6\downarrow \\
\hline
12 \\
\end{array}
$$

Because the decimal point is moved two units to the right in the divisor of 0.06 to turn it into the integer 6, it is also moved two units to the right in the dividend of 12.72 to make it 1,272. The result is 212, and remember that a division problem can always be checked by multiplying the answer by the divisor to see if the result is equal to the dividend.

Sometimes it is helpful to round answers that are in decimal form. First, find the place to which the rounding needs to be done. Then, look at the digit to the right of it. If that digit is 4 or less, the number in the place value to its left stays the same, and everything to its right becomes a zero. This process is known as **rounding down**. If that digit is 5 or higher, round up by increasing the place value to its left by 1, and every number to its right becomes a zero. If those zeros are in decimals, they can be dropped. For example, 0.145 rounded to the nearest hundredths place would be rounded up to 0.15, and 0.145 rounded to the nearest tenths place would be rounded down to 0.1.

Applying Estimation Strategies and Rounding Rules to Real-World Problems

Sometimes it is helpful to find an estimated answer to a problem rather than working out an exact answer. An estimation might be much quicker to find, and given the scenario, an estimation might be all that is required. For example, if Aria goes grocery shopping and has only a $100 bill to cover all of her purchases, it might be appropriate for her to estimate the total of the items she is purchasing to determine if she has enough money to cover them. Also, an estimation can help determine if an answer makes sense. For instance, if an answer in the 100s is expected, but the result is a fraction less than 1, something is probably wrong in the calculation.

The first type of estimation involves rounding. As mentioned, **rounding** consists of expressing a number in terms of the nearest decimal place like the tenth, hundredth, or thousandth place, or in terms of the nearest whole number unit like tens, hundreds, or thousands place. When rounding to a specific place value, look at the digit to the right of the place. If it is 5 or higher, round the number to its left up to the next value, and if it is 4 or lower, keep that number at the same value. For instance, 1,654.2674 rounded to the nearest thousand is 2,000, and the same number rounded to the nearest thousandth is 1,654.267. Rounding can be used in the scenario when grocery totals need to be estimated. Items can be rounded to the nearest dollar. For example, a can of corn that costs $0.79 can be rounded to $1.00, and then all other items can be rounded in a similar manner and added together.

When working with larger numbers, it might make more sense to round to higher place values. For example, when estimating the total value of a dealership's car inventory, it would make sense to round the car values to the nearest thousands place. The price of a car that is on sale for $15,654 can be estimated at $16,000. All other cars on the lot could be rounded in the same manner, and then their sum can be found. Depending on the situation, it might make sense to calculate an over-estimate. For example, to make sure Aria has enough money at the grocery store, rounding up every time for each item would ensure that she will have enough money when it comes time to pay. A $0.40 item rounded up to $1.00 would ensure that there is a dollar to cover that item. Traditional rounding rules would round $0.40 to $0, which does not make sense in this particular real-world setting. Aria might not have a dollar available at checkout to pay for that item if she uses traditional rounding. It is up to the customer to decide the best approach when estimating.

Estimating is also very helpful when working with measurements. Bryan is updating his kitchen and wants to retile the floor. Again, an over-measurement might be useful. Also, rounding to nearest half-unit might be helpful. For instance, one side of the kitchen might have an exact measurement of 14.32 feet, and the most useful measurement needed to buy tile could be estimating this quantity to be 14.5 feet. If the kitchen was rectangular and the other side measured 10.9 feet, Bryan might round the other side to 11 feet. Therefore, Bryan would find the total tile necessary according to the following area calculation:

$$4.5 \times 11 = 159.5 \text{ square feet}$$

To make sure he purchases enough tile, Bryan would probably want to purchase at least 160 square feet of tile. This is a scenario in which an estimation might be more useful than an exact calculation. Having more tile than necessary is better than having an exact amount, in case any tiles are broken or otherwise unusable.

Finally, estimation is helpful when exact answers are necessary. Consider a situation in which Sabina has many operations to perform on numbers with decimals, and she is allowed a calculator to find the result. Even though an exact result can be obtained with a calculator, there is always a possibility that Sabina could make an error while inputting the data. For example, she could miss a decimal place, or misuse a parenthesis, causing a problem with the actual order of operations. In this case, a quick estimation at the beginning would be helpful to make sure the final answer is given with the correct number of units.

Sabina has to find the exact total of 10 cars listed for sale at the dealership. Each price has two decimal places included to account for both dollars and cents. If one car is listed at $21,234.43 but Sabina incorrectly inputs into the calculator the price of $2,123.443, this error would throw off the final sum by almost $20,000. A quick estimation at the beginning, by rounding each price to the nearest thousands place and finding the sum of the prices, would give Sabina an amount to compare the exact amount to. This comparison would let Sabina see if an error was made in her exact calculation.

Practice Questions

Number Series

1. Examine this series: 4, 8, 12, 16 ...

What number should come next?
- a. 19
- (b.) 20
- c. 22
- d. 24

2. Examine this series: 0, 1, 1, 2, 3, 5, 8 ...

What number should come next?
- (a.) 9
- b. 10
- c. 11
- d. 13

3. Examine this series: 46, 43, 40, 37 ...

What number should come next?
- (a.) 34
- b. 33
- c. 32
- d. 31

4. Examine this series: 2, 4, 14, 16, 26, 28 ...

What number should come next?
- (a.) 30
- b. 32
- c. 38
- d. 40

5. Examine this series: 21, 27, 22, 28, 23, 29 ...

What number should come next?
- (a.) 24
- b. 25
- c. 30
- d. 31

6. Examine this series: 56, 49, 42 ...

What number should come next?
    a. 39
    b. 37
    c. 35
    d. 32

7. Examine this series: 6, 7, 9, 12, 16 ...

What number should come next?
    a. 18
    b. 19
    c. 20
    d. 21

8. Examine this series: 1, 4, 9, 16 ...

What number should come next?
    a. 21
    b. 25
    c. 32
    d. 36

9. Examine this series: 30, 22, 14, 6, ...

What number should come next?
    a. 1
    b. -2
    c. -3
    d. -6

10. Examine this series: 2, 6, 18, 54 ...

What number should come next?
    a. 72
    b. 108
    c. 162
    d. 208

11. Examine this series: 9, 12, 7, 10, 5, 8 ...

What number should come next?
    a. 3
    b. 6
    c. 14
    d. 15

12. Examine this series: 5, 15, 18, 54, 57, 171 ...

What number should come next?
   a. 173
   b. 174
   c. 177
   d. 522

13. Examine this series: 4, 8, 3, 6, 1, 2, -3 ...

What number should come next?
   a. -11
   b. -6
   c. -5
   d. -1

14. Examine this series: 1, 2, -4, -8, 16 ...

What number should come next?
   a. -32
   b. -24
   c. 24
   d. 32

15. Examine this series: 2, 4, 6, 12, 14, 28 ...

What number should come next?
   a. 30
   b. 32
   c. 56
   d. 60

16. Examine this series: 160, 80, 40 ...

What number should come next?
   a. 20
   b. 10
   c. 0
   d. -10

17. Examine this series: 1, 2, 4, 8, 10, 20 ...

What number should come next?
   a. 22
   b. 35
   c. 30
   d. 40

18. Examine this series: 2, 3, 6, 11, 18 ...

What number should come next?
   a. 25
   b. 26
   c. 27
   d. 28

Geometric Comparison

19. Examine I, II, and III and select the best answer.

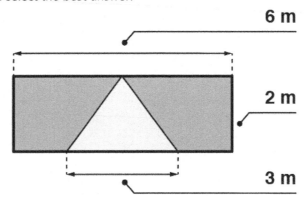

   I. The area of the dark gray shaded region
   II. 2 times the area of the light gray triangle
   III. One-half of the area of the entire rectangle
   a. I = II = III
   b. I > III > II
   c. II = III > I
   d. I > II = III

20. Examine I, II, and III and select the best answer.

An angle measures 45 degrees.

   I. The measure of a supplementary angle
   II. The measure of a complementary angle
   III. The measure of a bisected right angle
   a. II > III > I
   b. I = III > II
   c. II = III < I
   d. II > I > II

21. Examine I, II, and III and select the best answer.

The triangles in the following figure are similar.

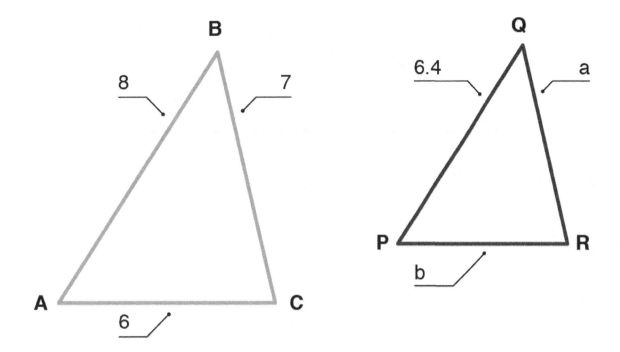

I. $\dfrac{a}{1.25}$

II. b

III. The perimeter of triangle ABC – the perimeter of triangle PQR

a. I > II > III

b. II > III > I

c. II > I > III

d. I = II = III

22. Examine I, II, and III and select the best answer.
    I. The value of *x* in the image below

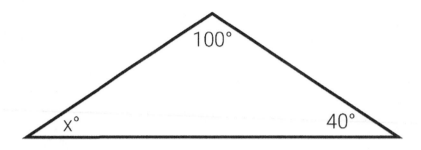

    II. The value of *x* in the image below

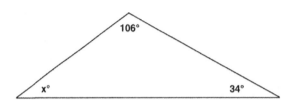

    III. The value of *x* in the image below

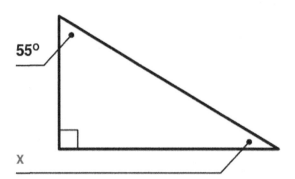

    a. I = II = III
    b. II > I > III
    c. I = II > III
    d. I = II < III

96

23. Examine I, II, and III and select the best answer.

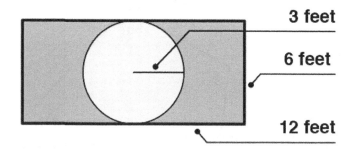

3 feet

6 feet

12 feet

    I. The area of the dark gray region
    II. 1.5 times the area of the circle
    III. One-half the area of the entire rectangle
a. I > II > III
b. II > I > III
c. II = III > I
d. I = II > III

24. Examine I, II, and III and select the best answer.

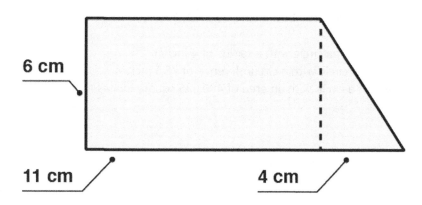

6 cm

11 cm

4 cm

    I. The area of the figure above
    II. 4.5 times the area of the triangle on the right
    III. The area of just the rectangle if its dimensions were each increased by 1 cm.
a. I = II > III
b. III > I > II
c. III = II > I
d. I = II = III

25. Examine I, II, and III and select the best answer.

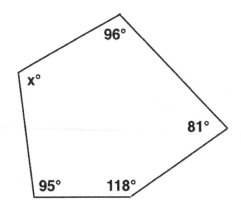

     I. The measure of angle *x*
     II. The measure of any angle in a regular pentagon
     III. The measure of any angle in a regular octagon
a. I > II > III
b. III > I > II
c. III > II > I
d. I > III > II

26. Examine I, II, and III and select the best answer. Use 3.14 for π.

     I. The circumference of a circle with a radius of 4 inches
     II. The diameter of a circle with a circumference of 78.5 inches
     III. The diameter of a circle with an area of 490.625 square inches
a. I = II = III
b. I = II > III
c. I > II = III
d. I > III > II

27. Examine I, II, and III and select the best answer.

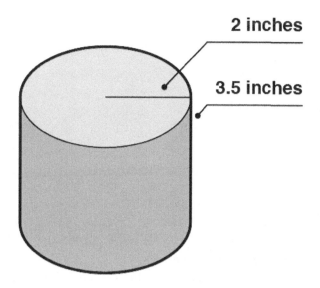

2 inches

3.5 inches

I. The volume of the cylinder above

II. 2 times the volume of a sphere with the same radius as the cylinder

III. The volume of a rectangular prism with the same height as the cylinder with a square base the length of the diameter.

a. II > I = III

b. II > III > I

c. I = II > III

d. II > I > III

Non-Geometric Comparison

28. Examine I, II, and III and select the best answer.

I. a, if $-8a + 4 + 9a = 35$

II. b, if $-2.5b + 5 - 6.5b = 18.5$

III. c, if $3(4.5 + 4c) = 16.5$

a. I > II > III

b. III < I < II

c. II < I = III

d. I > III > II

29. Examine I, II, and III and select the best answer.

      I. The result of dividing 24 by $\frac{8}{5}$

      II. The product of $\frac{14}{15}$ and $\frac{2}{5}$

      III. $6\frac{2}{3} - 2\frac{3}{4}$

a. I > II > III
b. III < I < II
c. II < I = III
d. I > III > II

30. Examine I, II, and III and select the best answer.

      I. The value of $y(9 - 7x^4 + 2x)$ when $x = -3$ and $y = 7$
      II. The value of $8n + 5n^3 + 16n^2$ when $n = 4$
      III. The value of $9k^2 - 6l^2 + 8k$ when $k = 12$ and $l = -8$

a. I > II > III
b. III < I < II
c. II < I = III
d. III > II > I

31. Examine I, II, and III and select the best answer.

Museum Visitors

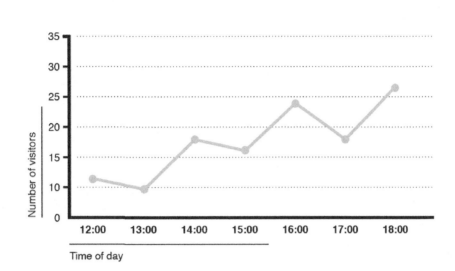

Time of day

I. The mean number of visitors for the first four hours
II. The median number of visitors for the full seven hours
III. The range of the number of visitors over the seven hours

a. II = III > I
b. II > III > I
c. II > I > III
d. II = I > III

32. Examine I, II, and III and select the best answer.

I. The percentage of math problems completed if 28 remain and there were 35 to start with
II. The percentage of students who live on campus if 40 students out of 200 live on campus
III. The value of $\frac{12}{60}$ converted to a percentage

a. I = II = III
b. III = II > I
c. III > I > II
d. I = II > III

33. Examine I, II, and III and select the best answer.

I. $8^2 - 3(4 + 5 \times 2)$
II. $(\sqrt{81} - \sqrt{49}) \times (72 \div 9)$
III. $3(8 \div 2^2)^3 + (5 \times 8) - 42$

a. I = II = III
b. III = II > I
c. III > I > II
d. I = III > II

34. Examine I, II, and III and select the best answer.

$$x + 3y = 18; -x - 4y = -25$$

       I. The value of x
       II. The value of y
       III. $x + y$
  a. III > II > I
  b. II > III > I
  c. III < II < I
  d. II < III < I

35. Examine I, II, and III and select the best answer.

$$-8x - 3y = 75; 4x + y = -33$$

       I. The value of x
       II. The value of y
       III. $1.5x$
  a. I> III > II
  b. I > II > III
  c. I > II = III
  d. II = III > I

Number Manipulation

36. What is $\frac{15}{60}$ converted to a percentage?
  a. 0.25
  b. 2.5%
  c. 25%
  d. 0.25%

37. Which of the following represents the correct sum of $\frac{14}{15}$ and $\frac{2}{5}$?
  a. $\frac{20}{15}$

  b. $\frac{4}{3}$

  c. $\frac{16}{20}$

  d. $\frac{4}{5}$

38. What is the product of $\frac{5}{9}$ and $\frac{7}{15}$?

  a. $\frac{7}{27}$

  b. $\frac{35}{135}$

  c. $\frac{5}{27}$

  d. $\frac{13}{45}$

39. What is the result of dividing 2 by $\frac{8}{5}$?

  a. $1\frac{1}{8}$

  b. $1\frac{2}{5}$

  c. $1\frac{1}{5}$

  d. $1\frac{1}{4}$

40. Subtract $\frac{5}{14}$ from $\frac{5}{24}$. Which of the following is the correct result?

  a. $\frac{25}{168}$

  b. 0

  c. $-\frac{25}{168}$

  d. $\frac{1}{10}$

41. Which of the following is a correct mathematical statement?

  a. $\frac{1}{3} < -\frac{4}{3}$

  b. $-\frac{1}{3} > \frac{4}{3}$

  c. $\frac{1}{3} > -\frac{4}{3}$

  d. $-\frac{1}{3} \geq \frac{4}{3}$

42. Which of the following is INCORRECT?

  a. $-\frac{1}{5} < \frac{4}{5}$

  b. $\frac{4}{5} > -\frac{1}{5}$

  c. $-\frac{1}{5} > \frac{4}{5}$

  d. $\frac{1}{5} > -\frac{4}{5}$

43. How many cases of cola can Lexi purchase if each case is $3.50 and she has $40?
   a. 10
   b. 12
   c. 11.4
   d. 11

44. A car manufacturer usually makes 15,412 SUVs, 25,815 station wagons, 50,412 sedans, 8,123 trucks, and 18,312 hybrids a month. About how many cars are manufactured each month?
   a. 120,000
   b. 200,000
   c. 300,000
   d. 12,000

45. Each year, a family goes to the grocery store every week and spends $105. About how much does the family spend annually on groceries?
   a. $10,000
   b. $50,000
   c. $500
   d. $5,000

46. A grocery store sold 48 bags of apples in one day, and 9 of the bags contained Granny Smith apples. The rest contained Red Delicious apples. What is the ratio of bags of Granny Smith to bags of Red Delicious apples that were sold?
   a. 48:9
   b. 39:9
   c. 9:48
   d. 9:39

47. If Oscar's bank account totaled $4,000 in March and $4,900 in June, what was the rate of change in his bank account over those three months?
   a. $900 a month
   b. $300 a month
   c. $4,900 a month
   d. $100 a month

48. Erin and Katie work at the same ice cream shop. Together, they always work less than 21 hours a week. In a week, if Katie worked two times as many hours as Erin, how many hours did Erin work?
   a. Less than 7 hours
   b. Less than or equal to 7 hours
   c. More than 7 hours
   d. Less than 8 hours

49. Which of the following is the correct decimal form of the fraction $\frac{14}{33}$ rounded to the nearest hundredth place?
   a. 42
   b. 0.42
   c. 0.424
   d. 0.140

50. Gina took an algebra test last Friday. There were 35 questions, and she answered 60% of them correctly. How many correct answers did she have?

    a. 35

    b. 20

    c. 21

    d. 25

51. Paul took a written driving test, and he got 12 of the questions correct. If he answered 75% of the questions correctly, how many problems were there in the test?

    a. 25

    b. 16

    c. 20

    d. 18

52. What is the solution to the equation $3(x + 2) = 14x - 5$?

    a. $x = 1$

    b. $x = 0$

    c. All real numbers

    d. There is no solution

Answer Explanations

1. B: The pattern in this series is that each number is a subsequent multiple of 4. Therefore, after 16, the next number in the series is 20.

2. D: The pattern in this series is that the two previous terms are added to get the next one, so:

$$5 + 8 = 13$$

3. A: The pattern in this series is that 3 is subtracted from the previous term to get the next one, so:

$$37 - 3 = 34$$

4. C: The pattern in this series is that there is an alternating pattern of adding 2 to the previous term and then adding 10 to the previous term, so 2 was added to 26 to get 8, which means we need to add 10 next, which gives us 38.

5. A: The pattern in this series is that there is an alternating pattern of alternating adding 6 and then subtracting 5, so 6 was added to 21 to get 27. Then, 5 was subtracted to get 22, 6 was added to get 28, 5 was subtracted for 23, 6 was added to get 29, so we must subtract 5 to get 24.

6. C: The pattern in this series is that 7 is subtracted from the previous term, so:

$$42 - 7 = 35$$

7. D: The pattern in this series is that you add one more than was added to get the previous term (add 1, then add 2, then add 3), so:

$$16 + 5 = 21$$

8. B: The terms in this series are found by taking the perfect square of the number of the term in the series (i.e., the third term is the perfect square of 3, which is 9). The fifth term, therefore, is 25.

9. B: The pattern in this series is that 8 is subtracted from the previous term, so:

$$6 - 8 = -2$$

10. C: The pattern in this series is that the previous term is multiplied by 3, so

$$54 \times 3 = 162$$

11. A: The pattern in this series is that the series alternates adding 3 and then subtracting 5 from the previous number, so 3 was added to 9 to get 12, then 5 was subtracted from 12 to get 7, then 3 was added to get 10, 5 was subtracted to get 5, 3 was added to get 8, and we need to subtract 5 to get 3.

12. B: The terms in this series are found by alternating multiplying the previous term by 3 then adding 3 to the previous term, so 171 was derived by multiplying 57 by 3. Therefore, next we add three to get 174.

13. B: The terms in this series are found by alternating multiplying the previous term by 2 then subtracting 5 from the previous term, so:

$$-3 \times 2 = -6$$

14. D: The terms in this series are found by alternating multiplying by positive 2 then negative 2.

$$1 \times 2 = 2$$

Then, $2 \times -2 = -4$.

Then $-4 \times 2 = -8$.

Then $-8 \times -2 = 16$.

Then $16 \times 2 = 32$.

15. A: The terms in this series are found by alternating multiplying by 2 then adding 2. The 28 was derived by multiplying 14 by 2, so next we add 2 to get 30.

16. A: Each term is half of the previous term, or the previous term divided by 2. Forty divided by 2 is 20.

17. A: Here, the pattern is alternating multiplying by 2 and then adding 2, so the next term is:

$$20 + 2 = 22$$

18. C: The difference between each term increases by 2 each time (first it's 1, then 3, then 5, then 7...). So:

$$18 + 9 = 27$$

19. D: The area of the dark shaded region is calculated in a few steps. First, the area of the rectangle is found using the formula:

$$A = length \times width = 6\,\text{m} \times 2\,\text{m} = 12\,\text{m}^2$$

Therefore, III is 6 m². Second, the area of the triangle is found using the formula:

$$A = \frac{1}{2} \times base \times height = \frac{1}{2} \times 3\,\text{m} \times 2\,\text{m} = 3\,\text{m}^2$$

Therefore, II is also 6 m². The last step is to take the rectangle area and subtract the triangle area. The area of the shaded region is:

$$A = 12 - 3 = 9\,\text{m}^2$$

Therefore, I > II = III.

20. C: For I, two supplementary angles sum to 180 degrees, so:

$$180 - 45 = 135\ degrees$$

That means that the supplementary angle has a measure of 135 degrees.

For II, two complementary angles have a sum of 90 degrees, so:

$$90 - 45 = 45\ degrees$$

Therefore, the complementary angle has a measure of 45 degrees.

Lastly, a right angle is 90 degrees, so bisecting that yields angles that are 45 degrees; thus, II = III < I.

21. C: Because the triangles are similar, the proportions:

$$\frac{8}{6.4} = \frac{6}{b}$$

and

$$\frac{8}{6.4} = \frac{7}{a}$$

can both be cross-multiplied and solved to obtain $a = 5.6$ and $b = 4.8$. Therefore, I is

$$\frac{5.6}{1.25} = 4.48 \text{ units}$$

For III, the perimeter of triangle ABC is

$$8 + 7 + 6 = 21 \text{ units}$$

while the perimeter of triangle PQR is:

$$6.4 + 5.6 + 4.8 = 16.8 \text{ units}$$

so $21 - 16.8 = 4.2$ units. Thus, II > I > III.

22. C: The sum of the angles in a triangle is 180 degrees. Therefore, to find x in I:

$$180 - (100 + 40) = 40°$$

In II:

$$180 - (106 + 34) = 40°$$

In III:

$$180 - (90 + 55) = 35°$$

Thus, I = II > III.

23. A: To find the area of the shaded region in the figure, determine the area of the whole figure. Then the area of the circle can be subtracted from the whole. The following formula shows the area of the outside rectangle:

$$A = 12 \times 6 = 72 \text{ ft}^2$$

Therefore, the area for III is 36 ft². The area of the inside circle can be found by the following formula:

$$A = \pi(3)^2 = 9\pi = 28.3 \text{ ft}^2$$

Thus, the area of II is 42.5 ft². As the shaded area is outside the circle, the area for the circle can be subtracted from the area of the rectangle to yield an area of 43.7 ft². That means I > II > III.

24. B: To find the area of the figure, break the figure into two shapes. The rectangle has dimensions 6 cm by 11 cm. The triangle has dimensions 6 cm by 4 cm. Plug the dimensions into the rectangle formula:

$$A = 6 \text{ cm} \times 11 \text{ cm} = 66 \text{ cm}^2$$

Multiplication yields an area of 66 cm². The triangle area can be found using the formula:

$$A = \frac{1}{2} \times 4 \text{ cm} \times 6 \text{ cm} = 12 \text{ cm}^2$$

Multiplication yields an area of 12 cm². Add the areas of the two shapes to find the total area of the figure, which is 78 cm².

For II, the area of the triangle was 12 cm², so:

$$4.5 \times 12 \text{ cm}^2 = 54 \text{ cm}^2$$

For III, the area of the new rectangle would be:

$$A = 7 \text{ cm} \times 12 \text{ cm} = 84 \text{ cm}^2$$

Multiplication yields an area of 56 cm². Thus, III > I > II.

25. D: To find the sum of the interior angles in a polygon with n sides, the expression $(n - 2) \times 180°$ is used. Since the polygon in the figure has 5 sides, then an equation to find the total degrees of the interior angles can be formed as follows:

$$(5 - 2) \times 180° = 540°$$

To find the measurement of angle x, we add the angles given and subtract that total from 540 degrees.

$$96° + 81° + 118° + 95° = 390°$$

$$540° - 390° = 150°$$

The measure of any of the angles in a regular pentagon is found by dividing the sum of the interior angles (540 degrees) by 5; this yields 108 degrees. The measure of any of the angles in a regular octagon is found by dividing the sum of the interior angles by 8. Using $(n - 2) \times 180°$ when $n = 8$ gives:

$$6 \times 180° = 1080°$$

Dividing this by 8 yields 135 degrees.

Thus, I > III > II.

26. C: The formula for the circumference of a circle is $C = 2\pi r$ or $C = \pi d$, since the diameter is two times the length of the radius. Thus, the circumference for the circle in I is:

$$C = 2\pi 4 = 8\pi$$

Using 3.14 for π yields 25.12 inches. For II, we can use the formula in reverse. $78.5 = \pi d$. Dividing both sides by 3.14 gives $d = 25$ inches. So, II is slightly smaller than I. For III, the area of a circle is calculated via the formula $A = \pi r^2$, so $490.625 = \pi r^2$. Dividing both sides by 3.14 gives $156.25 = r^2$, so $r = 12.5$.

Because the radius is half the length of the diameter, the diameter of the circle is 25 inches. Thus I > II = III.

27. B: The volume for a cylinder is found by using the formula:

$$V = \pi r^2 h$$

$$\pi(2^2) \times 3.5 = 43.96 \text{ in}^3$$

The volume for a sphere is found by using the formula:

$$V = \frac{4}{3}\pi r^3$$

$$V = \frac{4}{3}\pi(2^3) = \frac{4}{3}8\pi = 33.51 \text{ in}^3$$

To double that, we multiply $33.51 \times 2 = 67.02 \text{ in}^3$, so II >I.

The volume for a rectangular prism is found by using the formula:

$$V = lwh$$

$$(3.5)(4)(4) = 56 \text{ in}^3$$

Note that the diameter is twice the radius, so 4 inches was used as the side length for the rectangular prism.

Now, we can see that II > III > I.

28. D: For I, we need to simplify by combining like terms:

$$-8a + 4 + 9a = 35$$
$$a + 4 = 35$$
$$a = 31$$

For II, we need to simplify by combining like terms:

$$-2.5b + 5 - 6.5b = 18.5$$
$$5 - 9b = 18.5$$
$$-9b = 13.5$$
$$b = -1.5$$

For III, we need to distribute and then simplify:

$$3(4.5 + 4c) = 16.5$$
$$13.5 + 12c = 16.5$$
$$12c = 3$$
$$c = 0.25$$

Therefore, I > III > II.

29. D: For I, division is completed by multiplying times the reciprocal. Therefore:

$$24 \div \frac{8}{5} = \frac{24}{1} \times \frac{5}{8}$$

$$\frac{3 \times 8}{1} \times \frac{5}{8} = \frac{15}{1} = 15$$

II is:

$$\frac{14}{15} \times \frac{2}{5} = \frac{28}{75}$$

III is:

$$6\frac{2}{3} - 2\frac{3}{4} = 6\frac{8}{12} - 2\frac{9}{12}$$

$$5\frac{20}{12} - 2\frac{9}{12} = 3\frac{11}{12}$$

Therefore, I > III > II.

30. D: For I, each instance of x is replaced with -3, and each instance of y is replaced with 7 to get

$$y(9 - 7x^4 + 2x)$$

$$7\left(9 - 7(-3)^4 + 2(-3)\right)$$

$$7(9 - 567 - 6)$$

$$7(-564)$$

$$-3{,}948$$

For II, each instance of n is replaced with 4 to get:

$$8n + 5n^3 + 16n^2$$

$$8(4) + 5(4)^3 + 16(4)^2$$

$$32 + 320 + 256$$

$$608$$

For III, each instance of k is replaced with 12, and each instance of l is replaced with -8 to get

$$9k^2 - 6l^2 + 8k$$

$$9(12)^2 - 6(-8)^2 + 8(12)$$

$$1{,}296 - 384 + 96$$

$$1{,}008$$

Therefore, III > II > I.

31. B: The mean for the number of visitors during the first 4 hours is 14. The mean is found by calculating the average for the four hours. Adding up the total number of visitors during those hours gives:

$$12 + 10 + 18 + 16 = 56$$

Then $56 \div 4 = 14$.

The median of the total seven hours can be found by arranging the numbers in order (10, 12, 16, 18, 18, 24, 26) and finding the number in the middle, which is 18. The range is the difference in the highest number of visitors (26) and the fewest (10), which is 16. Thus, II > III > I.

32. A: For I, if 28 remain and there 35 questions, 7 questions have been completed, which in decimal form is $\frac{7}{35} = 0.2$. Then, to convert to a percentage, move the decimal point two units to the right and add the percentage symbol. The result is 20%.

For II, if 40 students out of 200 live on campus, then the corresponding fraction to represent this situation is $\frac{40}{200}$. This reduces to $\frac{1}{5}$ and can be converted to a percentage of 20%.

For III, the fraction $\frac{12}{60}$ can be reduced to $\frac{1}{5}$, in lowest terms. Then, it must be converted to a decimal. Dividing 1 by 5 results in 0.2. Then, to convert to a percentage, move the decimal point two units to the right and add the percentage symbol. The result is 20%.

33. D: Using order of operations, the expression for I can be solved as follows:

$$8^2 - 3(4 + 5 \times 2)$$

$$64 - 3(14) = 64 - 42 = 22$$

Using order of operations, the expression for II can be solved as follows:

$$\left(\sqrt{81} - \sqrt{49}\right) \times (72 \div 9)$$

$$(9 - 7) \times (8)$$

$$2 \times 8 = 16$$

Using order of operations, the expression for III can be solved as follows:

$$3(8 \div 2^2)^3 + (5 \times 8) - 42$$

$$3(8 \div 4)^3 + 40 - 42$$

$$3(2)^3 - 2$$

$$3(8) - 2$$

$$24 - 2$$

$$22$$

Therefore, I = III > II.

34. B: This is a system of equations:

$$x + 3y = 18; -x - 4y = -25$$

We can add these two equations to cancel the x term:

$$x + 3y + (-x - 4y) = 18 + (-25)$$

$$x + 3y - x - 4y = 18 - 25$$

$$-y = -7$$

$$y = 7$$

Next, we can plug in our calculated value for y into either equation to find x:

$$x + 3y = 18$$

$$x + 3(7) = 18$$

$$x = -3$$

Therefore, $x + y = 4$, so II > III > I.

35. C: This system of equations can be solved by substitution.

$$-8x - 3y = 75; 4x + y = -33$$

We can manipulate the second equation to be:

$$y = -33 - 4x$$

Then, we can use this value of y in our first equation:

$$-8x - 3(-33 - 4x) = 75$$

$$-8x + 99 + 12x = 75$$

Next, collect like terms to simplify:

$$4x = -24$$

$$x = -6$$

Now, we can use this value of x to find y:

$$y = -33 - 4(-6)$$

$$y = -33 + 24$$

$$y = -9$$

Therefore, $1.5x = -9$, so I > II = III.

36. C: The fraction $\frac{15}{60}$ can be reduced to $\frac{1}{4}$, which puts the fraction in lowest terms. First, it must be converted to a decimal. Dividing 1 by 4 results in 0.25. Then, to convert to a percentage, move the decimal point two units to the right and add the percentage symbol. The result is 25%.

37. B: Common denominators must be used. The LCD is 15, and $\frac{2}{5} = \frac{6}{15}$. Therefore:

$$\frac{14}{15} + \frac{6}{15} = \frac{20}{15}$$

and in lowest terms, the answer is $\frac{4}{3}$. A common factor of 5 was divided out of both the numerator and denominator.

38. A: A product is found by multiplying.

$$\frac{5}{9} \times \frac{7}{15}$$

$$\frac{35}{135}$$

Simplify by dividing 5 out of the numerator and denominator:

$$\frac{7}{27}$$

39. D: Division is completed by multiplying by the reciprocal. Therefore:

$$2 \div \frac{8}{5} = \frac{2}{1} \times \frac{5}{8}$$

$$\frac{10}{8} = 1\frac{1}{4}$$

40. C: Common denominators must be used. The LCD is 168, so each fraction must be converted to have 168 as the denominator.

$$\frac{5}{24} - \frac{5}{14} = \frac{5}{24} \times \frac{7}{7} - \frac{5}{14} \times \frac{12}{12}$$

$$\frac{35}{168} - \frac{60}{168} = -\frac{25}{168}$$

41. C: The correct mathematical statement is the one in which the smaller of the two numbers is on the "less than" side of the inequality symbol. It is written in Choice C that $\frac{1}{3} > -\frac{4}{3}$, which is the same as:

$$-\frac{4}{3} < \frac{1}{3}$$

a correct statement.

42. C: $-\frac{1}{5} > \frac{4}{5}$ is incorrect. The expression on the left is negative, which means that it is smaller than the expression on the right. As it is written, the inequality states that the expression on the left is greater than the expression on the right, which is not true.

43. D: This is a one-step real-world application problem. The unknown quantity is the number of cases of cola to be purchased. Let x be equal to this amount. Because each case costs $3.50, the total number of cases multiplied by $3.50 must equal $40. This translates to the mathematical equation $3.5x = 40$. Divide both sides by 3.5 to obtain $x = 11.4286$, which has been rounded to four decimal places. Because cases are sold whole, and there is not enough money to purchase 12 cases, 11 cases is the correct answer.

44. A: Rounding can be used to find the best approximation. All of the values can be rounded to the nearest thousand. 15,412 SUVs can be rounded to 15,000. 25,815 station wagons can be rounded to 26,000. 50,412 sedans can be rounded to 50,000. 8,123 trucks can be rounded to 8,000. Finally, 18,312 hybrids can be rounded to 18,000. The sum of the rounded values is 117,000, which is closest to 120,000.

45. D: There are 52 weeks in a year, and if the family spends $105 each week, that amount is close to $100. A good approximation is $100 a week for 50 weeks, which is found through the product:

$$50 \times 100 = \$5,000$$

46. D: There were 48 total bags of apples sold. If 9 bags were Granny Smith and the rest were Red Delicious, then $48 - 9 = 39$ bags were Red Delicious. Therefore, the ratio of Granny Smith to Red Delicious is 9:39.

47. B: The average rate of change is found by calculating the difference in dollars over the elapsed time. Therefore, the rate of change is equal to:

$$(\$4,900 - \$4,000) \div 3 \text{ months}$$

which is equal to $900 ÷ 3 or $300 per month.

48. A: Let x be the unknown, the number of hours Erin can work. We know Katie works $2x$, and the sum of all hours is less than 21. Therefore:

$$x + 2x < 21$$

which simplifies into $3x < 21$. Solving this results in the inequality $x < 7$ after dividing both sides by 3. Therefore, Erin worked less than 7 hours.

49. B: If a calculator is used, divide 33 into 14 and keep two decimal places. If a calculator is not used, multiply both the numerator and denominator times 3. This results in the fraction $\frac{42}{99}$, and hence a decimal of 0.42.

50. C: Gina answered 60% of 35 questions correctly; 60% can be expressed as the decimal 0.60. Therefore, she answered $0.60 \times 35 = 21$ questions correctly.

51. B: The unknown quantity is the number of total questions on the test. Let x be equal to this unknown quantity. Therefore, $0.75x = 12$. Divide both sides by 0.75 to obtain $x = 16$.

52. A: First, the distributive property must be used on the left side. This results in $3x + 6 = 14x - 5$. The addition property is then used to add 5 to both sides, and then to subtract $3x$ from both sides, resulting in $11 = 11x$. Finally, the multiplication property is used to divide each side by 11. Therefore, $x = 1$ is the solution.

Reading

Comprehension

Summarizing a Text

An important skill is the ability to read a complex text and then reduce its length and complexity by focusing on the key events and details. A **summary** is a shortened version of the original text, written by the reader in their own words. The summary should be shorter than the original text, and it must be thoughtfully formed to include critical points from the original text.

In order to effectively summarize a complex text, it's necessary to understand the original source and identify the major points covered. It may be helpful to outline the original text to get the big picture and avoid getting bogged down in the minor details. For example, a summary wouldn't include a statistic from the original source unless it was the major focus of the text. It's also important for readers to use their own words, yet retain the original meaning of the passage. The key to a good summary is emphasizing the main idea without changing the focus of the original information.

The more complex a text, the more difficult it can be to summarize. Readers must evaluate all points from the original source and then filter out what they feel are the less necessary details. Only the essential ideas should remain. The summary often mirrors the original text's organizational structure. For example, in a problem-solution text structure, the author typically presents readers with a problem and then develops solutions through the course of the text. An effective summary would likely retain this general structure, rephrasing the problem and then reporting the most useful or plausible solutions.

Paraphrasing is somewhat similar to summarizing. It calls for the reader to take a small part of the passage and list or describe its main points. Paraphrasing is more than rewording the original passage, though. Like a summary, it should be written in the reader's own words, while still retaining the meaning of the original source. The main difference between summarizing and paraphrasing is that a summary would be appropriate for a much larger text, while a paraphrase might focus on just a few lines of text. Effective paraphrasing will indicate an understanding of the original source, yet still help the readers expand on their interpretation. A paraphrase should neither add new information nor remove essential facts that change the meaning of the source.

Identifying the Topic, Main Idea, and Supporting Details

The **topic** of a text is the general subject matter. Text topics can usually be expressed in one word, or a few words at most. Additionally, readers should ask themselves what point the author is trying to make. This point is the **main idea** of the text—the one thing the author wants readers to know about the topic. Once the author has established the main idea, he or she will support the main idea with supporting details. **Supporting details** are evidence that support the main idea and include personal testimonies, examples, or statistics.

One analogy for these components and their relationships is that a text is like a well-designed house. The topic is the roof, covering all rooms. The main idea is the frame. The supporting details are the various rooms. To identify the topic of a text, readers can ask themselves what or who the author is writing about in the paragraph. To locate the main idea, readers can ask themselves what one idea the author wants readers to know about the topic. To identify supporting details, readers can put the main idea into question form and ask, "what does the author use to prove or explain their main idea?"

Let's look at an example. An author is writing an essay about the Amazon rainforest and trying to convince the audience that more funding should go into protecting the area from deforestation. The author makes the argument stronger by including evidence of the benefits of the rainforest: it provides habitats to a variety of species, it provides much of the earth's oxygen which in turn cleans the atmosphere, and it is the home to medicinal plants that are useful against some of the world's deadliest diseases. Here is an outline of the essay looking at topic, main idea, and supporting details:

- Topic: Amazon rainforest
- Main Idea: The Amazon rainforest should receive more funding to protect it from deforestation.
- Supporting Details:
  1. It provides habitats to a variety of species
  2. It provides much of the earth's oxygen which in turn cleans the atmosphere
  3. It is home to medicinal plants that are useful against some of the deadliest diseases.

Notice that the topic of the essay is listed in a few key words: "Amazon rainforest." The main idea tells us what about the topic is important: that the topic should be funded to prevent deforestation. Finally, the supporting details are what author relies on to convince the audience to act or to believe in the truth of the main idea.

Theme of a Text
The **theme** of a text is the central idea the author communicates. Whereas the topic of a passage of text may be concrete in nature, by contrast the theme is always conceptual. For example, while the topic of Mark Twain's novel *The Adventures of Huckleberry Finn* might be described as something like the coming-of-age experiences of a poor, illiterate, functionally orphaned boy around and on the Mississippi River in 19th-century Missouri, one theme of the book might be that human beings are corrupted by society. Another might be that slavery and "civilized" society itself are hypocritical. Whereas the main idea in a text is the most important single point that the author wants to make, the theme is the concept or view around which the author centers the text.

Throughout time, humans have told stories with similar themes. Some themes are universal across time, space, and culture. These include themes of the individual as a hero, conflicts of the individual against nature, the individual against society, change vs. tradition, the circle of life, coming-of-age, and the complexities of love. Themes involving war and peace have featured prominently in diverse works, like Homer's *Iliad*, Tolstoy's *War and Peace* (1869), Stephen Crane's *The Red Badge of Courage* (1895), Hemingway's *A Farewell to Arms* (1929), and Margaret Mitchell's *Gone with the Wind* (1936). Another universal literary theme is that of the quest. These appear in folklore from countries and cultures worldwide, including the Gilgamesh Epic, Arthurian legend's Holy Grail quest, Virgil's *Aeneid*, Homer's *Odyssey*, and the *Argonautica*. Cervantes' *Don Quixote* is a parody of chivalric quests. J.R.R. Tolkien's *The Lord of the Rings* trilogy (1954) also features a quest.

One instance of similar themes across cultures is when those cultures are in countries that are geographically close to each other. For example, a folklore story of a rabbit in the moon using a mortar and pestle is shared among China, Japan, Korea, and Thailand—making medicine in China, making rice cakes in Japan and Korea, and hulling rice in Thailand. Another instance is when cultures are more distant geographically, but their languages are related. For example, East Turkestan's Uighurs and people in Turkey share tales of folk hero Effendi Nasreddin Hodja. Another instance, which may either be called cultural diffusion or simply reflect commonalities in the human imagination, involves shared themes among geographically- and linguistically-different cultures: both Cameroon's and Greece's folklore tell of centaurs; Cameroon, India, Malaysia, Thailand, and Japan, of mermaids; Brazil, Peru, China, Japan,

Malaysia, Indonesia, and Cameroon, of underwater civilizations; and China, Japan, Thailand, Vietnam, Malaysia, Brazil, and Peru, of shape-shifters.

Two prevalent literary themes are love and friendship, which can end happily, sadly, or both. William Shakespeare's *Romeo and Juliet*, Emily Brontë's *Wuthering Heights*, Leo Tolstoy's *Anna Karenina*, and both *Pride and Prejudice* and *Sense and Sensibility* by Jane Austen are famous examples. Another theme recurring in popular literature is of revenge, an old theme in dramatic literature, e.g. Elizabethans Thomas Kyd's *The Spanish Tragedy* and Thomas Middleton's *The Revenger's Tragedy*. Some more well-known instances include Shakespeare's tragedies *Hamlet* and *Macbeth*, Alexandre Dumas' *The Count of Monte Cristo*, John Grisham's *A Time to Kill*, and Stieg Larsson's *The Girl Who Kicked the Hornet's Nest*.

Identifying the Author's Purpose in a Given Text

Authors may have many purposes for writing a specific text. An **author's purpose** may be to try and convince readers to agree with their position on a subject, to impart information, or to entertain. Other writers are motivated to write from a desire to express their own feelings. Authors' purposes are their reasons for writing something. A single author may have one overriding purpose for writing or multiple reasons. An author may explicitly state their intention in the text, or the reader may need to infer that intention. Those who read reflectively benefit from identifying the purpose because it enables them to analyze information in the text. By knowing why the author wrote the text, readers can glean ideas for how to approach it.

The following is a list of questions readers can ask in order to discern an author's purpose for writing a text:

- From the title of the text, why do you think the author wrote it?
- Was the purpose of the text to give information to readers?
- Did the author want to describe an event, issue, or individual?
- Was it written to express emotions and thoughts?
- Did the author want to convince readers to consider a particular issue?
- Was the author primarily motivated to write the text to entertain?
- Why do you think the author wrote this text from a certain point of view?
- What is your response to the text as a reader?
- Did the author state their purpose for writing it?

Students should read to interpret information rather than simply content themselves with roles as text consumers. Being able to identify an author's purpose efficiently improves reading comprehension, develops critical thinking, and makes students more likely to consider issues in depth before accepting writer viewpoints. Authors of fiction frequently write to entertain readers. Another purpose for writing fiction is making a political statement; for example, Jonathan Swift wrote "A Modest Proposal" (1729) as a political satire. Another purpose for writing fiction as well as nonfiction is to persuade readers to take some action or further a particular cause. Fiction authors and poets both frequently write to evoke certain moods; for example, Edgar Allan Poe wrote novels, short stories, and poems that evoke moods of gloom, guilt, terror, and dread. Another purpose of poets is evoking certain emotions: love is popular, as in Shakespeare's sonnets and numerous others. In "The Waste Land" (1922), T.S. Eliot evokes society's alienation, disaffection, sterility, and fragmentation.

Authors seldom directly state their purposes in texts. Some students may be confronted with nonfiction texts such as biographies, histories, magazine and newspaper articles, and instruction manuals, among others. To identify the purpose in nonfiction texts, students can ask the following questions:

- Is the author trying to teach something?
- Is the author trying to persuade the reader?
- Is the author imparting factual information only?
- Is this a reliable source?
- Does the author have some kind of hidden agenda?

To apply author purpose in nonfictional passages, students can also analyze sentence structure, word choice, and transitions to answer the aforementioned questions and to make inferences. For example, authors wanting to convince readers to view a topic negatively often choose words with negative connotations.

Narrative Writing

Narrative writing tells a story. The most prominent examples of narrative writing are fictional novels. Here are some examples:

- Mark Twain's *The Adventures of Tom Sawyer* and *The Adventures of Huckleberry Finn*

- Victor Hugo's *Les Misérables*

- Charles Dickens' *Great Expectations, David Copperfield*, and *A Tale of Two Cities*

- Jane Austen's *Northanger Abbey, Mansfield Park, Pride and Prejudice, Sense and Sensibility*, and *Emma*

- Toni Morrison's *Beloved, The Bluest Eye*, and *Song of Solomon*

- Gabriel García Márquez's *One Hundred Years of Solitude* and *Love in the Time of Cholera*

Some nonfiction works are also written in narrative form. For example, some authors choose a narrative style to convey factual information about a topic, such as a specific animal, country, geographic region, and scientific or natural phenomenon.

Since narrative is the type of writing that tells a story, it must be told by someone, who is the narrator. The narrator may be a fictional character telling the story from their own viewpoint. This narrator uses the first person (*I, me, my, mine* and *we, us, our,* and *ours*). The narrator may simply be the author; for example, when Louisa May Alcott writes "Dear reader" in *Little Women*, she (the author) addresses us as readers. Many novels are typically told in third person, referring to the characters as *he, she, they,* or *them*. Another more common technique is the omniscient narrator; i.e. the story is told by an unidentified individual who sees and knows everything about the events and characters—not only their externalized actions, but also their internalized feelings and thoughts. Second person, i.e. writing the story by addressing readers as "you" throughout, is less frequently used.

Expository Writing

Expository writing is also known as informational writing. Its purpose is not to tell a story as in narrative writing, to paint a picture as in descriptive writing, or to persuade readers to agree with something as in argumentative writing. Rather, its point is to communicate information to the reader. As such, the point of view of the author will necessarily be more objective. Whereas other types of writing appeal to the reader's emotions, appeal to the reader's reason by using logic, or use subjective descriptions to sway the reader's opinion or thinking, expository writing seeks to do none of these but simply to provide facts, evidence, observations, and objective descriptions of the subject matter. Some examples of expository writing include research reports, journal articles, articles and books about historical events or periods, academic subject textbooks, news articles and other factual journalistic reports, essays, how-to articles, and user instruction manuals.

Technical Writing

Technical writing is similar to expository writing in that it is factual, objective, and intended to provide information to the reader. Indeed, it may even be considered a subcategory of expository writing. However, technical writing differs from expository writing in that (1) it is specific to a particular field, discipline, or subject; and (2) it uses the specific technical terminology that belongs only to that area. Writing that uses technical terms is intended only for an audience familiar with those terms. A primary example of technical writing today is writing related to computer programming and use.

Persuasive Writing

Persuasive writing is intended to persuade the reader to agree with the author's position. It is also known as argumentative writing. Some writers may be responding to other writers' arguments, in which case they make reference to those authors or text and then disagree with them. However, another common technique is for the author to anticipate opposing viewpoints in general, both from other authors and from the author's own readers. The author brings up these opposing viewpoints, and then refutes them before they can even be raised, strengthening the author's argument. Writers persuade readers by appealing to their reason, which Aristotle called **logos**; appealing to emotion, which Aristotle called **pathos**; or appealing to readers based on the author's character and credibility, which Aristotle called **ethos**.

The Organization and Structure of a Text

Comparison-Contrast Text Structure

Comparison-contrast text structure identifies similarities between two or more things or identifies differences between two or more things. Authors typically employ both to illustrate relationships between things by highlighting their commonalities and deviations. For example, a writer might compare Windows and Linux as operating systems, and contrast Linux as free and open-source vs. Windows™ as proprietary. When writing an essay, sometimes it is useful to create an image of the two objects or events you are comparing or contrasting. **Venn diagrams** are useful because they show the differences as well as the similarities between two things. Once you've seen the similarities and differences on paper, it might be helpful to create an outline of the essay with both comparison and contrast. Every outline will look different, because every two or more things will have a different number of comparisons and contrasts. Say you are trying to compare and contrast carrots with sweet potatoes.

Here is an example of a compare/contrast outline using those topics:

- Introduction: Share why you are comparing and contrasting the foods. Give the thesis statement.
- Body paragraph 1: Sweet potatoes and carrots are both root vegetables (similarity)
- Body paragraph 2: Sweet potatoes and carrots are both orange (similarity)
- Body paragraph 3: Sweet potatoes and carrots have different nutritional components (difference)
- Conclusion: Restate the purpose of your comparison/contrast essay.

Of course, if there is only one similarity between your topics and two differences, you will want to rearrange your outline. Always tailor your essay to what works best with your topic.

Descriptive Text Structure

Description can be both a type of text structure and a type of text. Some texts are descriptive throughout entire books. For example, a book may describe the geography of a certain country, state, or region, or tell readers all about dolphins by describing many of their characteristics. Many other texts are not descriptive throughout, but use descriptive passages within the overall text. The following are a few examples of **descriptive text structure**:

- When the author describes a character in a novel
- When the author sets the scene for an event by describing the setting
- When a biographer describes the personality and behaviors of a real-life individual
- When a historian describes the details of a particular battle within a book about a specific war
- When a travel writer describes the climate, people, foods, and/or customs of a certain place

A hallmark of description is using sensory details, painting a vivid picture so readers can imagine it almost as if they were experiencing it personally.

Cause and Effect Text Structure

When using cause and effect to extrapolate meaning from text, readers must determine the cause when the author only communicates effects. For example, if a description of a child eating an ice cream cone includes details like beads of sweat forming on the child's face and the ice cream dripping down her hand faster than she can lick it off, the reader can infer or conclude it must be hot outside. A useful technique for making such decisions is wording them in "If...then" form, e.g. "If the child is perspiring and the ice cream melting, it may be a hot day." **Cause and effect text structure** explains why certain events or actions resulted in particular outcomes. For example, an author might describe America's historical large flocks of dodo birds, the fact that gunshots did not startle/frighten dodos, and that because dodos did not flee, settlers killed whole flocks in one hunting session, explaining how the dodo was hunted into extinction.

Recognizing Events in a Sequence

Sequence structure is the order of events in which a story or information is presented to the audience. Sometimes the text will be presented in chronological order, or sometimes it will be presented by displaying the most recent information first, then moving backwards in time. The sequence structure depends on the author, the context, and the audience. The structure of a text also depends on the genre in which the text is written. Is it literary fiction? Is it a magazine article? Is it instructions for how to complete a certain task? Different genres will have different purposes for switching up the sequence of their writing.

Narrative Structure

The structure presented in literary fiction is also known as **narrative structure.** Narrative structure is the foundation on which the text moves. The basic ways for moving the text along are in the plot and the setting. The **plot** is the sequence of events in the narrative that move the text forward through cause and effect. The **setting** of a story is the place or time period in which the story takes place. Narrative structure has two main categories: linear and nonlinear.

Linear narrative is a narrative told in chronological order. Traditional linear narratives will follow the plot diagram below depicting the narrative arc. The narrative arc consists of the exposition, conflict, rising action, climax, falling action, and resolution.

- Exposition: The **exposition** is in the beginning of a narrative and introduces the characters, setting, and background information of the story. The importance of the exposition lies in its framing of the upcoming narrative. Exposition literally means "a showing forth" in Latin.

- Conflict: The **conflict**, in a traditional narrative, is presented toward the beginning of the story after the audience becomes familiar with the characters and setting. The conflict is a single instance between characters, nature, or the self, in which the central character is forced to make a decision or move forward with some kind of action. The conflict presents something for the main character, or protagonist, to overcome.

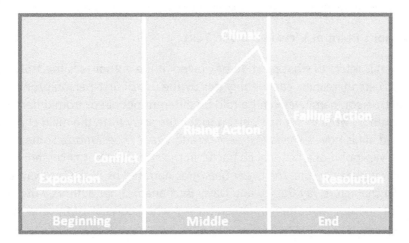

- Rising Action: The **rising action** is the part of the story that leads into the climax. The rising action will feature the development of characters and plot, and creates the tension and suspense that eventually lead to the climax.

- Climax: The **climax** is the part of the story where the tension produced in the rising action culminates. The climax is the peak of the story. In a traditional structure, everything before the climax builds up to it, and everything after the climax falls from it. It is the height of the narrative and is usually either the most exciting part of the story, or is marked by some turning point in the character's journey.

- Falling Action: The **falling action** happens as a result of the climax. Characters continue to develop, although there is a wrapping up of loose ends here. The falling action leads to the resolution.

- Resolution: The **resolution** is where the story comes to an end and usually leaves the reader with the satisfaction of knowing what happened within the story and why. However, stories do not always end in this fashion. Sometimes readers can be confused or frustrated at the end from the lack of information or the absence of a happy ending.

A **nonlinear narrative** deviates from the traditional narrative in that it does not always follow the traditional plot structure of the narrative arc. Nonlinear narratives may include structures that are disjointed, circular, or disruptive, in the sense that they do not follow chronological order, but rather a nontraditional order of structure. *In medias res* is an example of a structure that predates the linear narrative. **In medias res** is Latin for "in the middle of things," which is how many ancient texts, especially epic poems, began their story, such as Homer's *Iliad*. Instead of having a clear exposition with a full development of characters, they would begin right in the middle of the action.

Modernist texts in the late nineteenth and early twentieth century are known for their experimentation with disjointed narratives, moving away from traditional linear narrative. **Disjointed narratives** are depicted in novels like *Catch 22*, where the author, Joseph Heller, structures the narrative based on free association of ideas rather than chronology. Another nonlinear narrative can be seen in the novel *Wuthering Heights*, written by Emily Bronte, which disrupts the chronological order by being told retrospectively after the first chapter. It seems that there are two narratives in *Wuthering Heights* working at the same time: a present narrative as well as a past narrative. Authors employ disrupting narratives for various reasons; some use it to create situational irony for the readers, while some use it to create a certain effect in the reader, such as excitement, or even a feeling of discomfort or fear.

Evaluating the Author's Point of View in a Given Text

Point of view in a text is refers to what position or viewpoint the author tells the story. When a writer tells a story using the first person, readers can identify this by the use of first-person pronouns, like *I, me, we, us,* etc. However, first-person narratives can be told by different people or from different points of view. For example, some authors write in the first person to tell the story from the main character's viewpoint, as Charles Dickens did in his novels *David Copperfield* and *Great Expectations*. Some authors write in the first person from the viewpoint of a fictional character in the story, but not necessarily the main character. For example, F. Scott Fitzgerald wrote *The Great Gatsby* as narrated by Nick Carraway, a character in the story, about the main characters, Jay Gatsby and Daisy Buchanan. Other authors write in the first person, but as the omniscient narrator—an often unnamed person who knows all of the characters' inner thoughts and feelings. Writing in first person as oneself is more common in nonfiction.

Third Person

The **third-person point of view** is probably the most prevalent voice used in fictional literature. While some authors tell stories from the point of view and in the voice of a fictional character using the first person, it is a more common practice to describe the actions, thoughts, and feelings of fictional characters in the third person using *he, him, she, her, they, them,* etc.

Although plot and character development are both necessary and possible when writing narrative from a first-person point of view, they are also more difficult, particularly for new writers and those who find it unnatural or uncomfortable to write from that perspective. Therefore, writing experts advise beginning writers to start out writing in the third person. A big advantage of third-person narration is that the writer can describe the thoughts, feelings, and motivations of every character in a story, which is not possible for the first-person narrator. Third-person narrative can impart information to readers that the characters do not know. On the other hand, beginning writers often regard using the third-person point of view as more difficult because they must write about the feelings and thoughts of every character, rather than only about those of the protagonist.

Second Person

Second-person point of view addresses someone else as "you." In novels and other fictional works, the second person is the narrative voice most seldom used. The primary reason for this is that it often reads in an awkward manner, which prevents readers from being drawn into the fictional world of the novel. The second person is more often used in informational text, especially in how-to manuals, guides, and other instructions.

First Person

First-person point of view uses pronouns such as *I, me, we, my, us,* and *our.* Some writers naturally find it easier to tell stories from their own points of view, so writing in the first person offers advantages for them. The first-person voice is better for interpreting the world from a single viewpoint, and for enabling reader immersion in one protagonist's experiences. However, others find it difficult to use the first-person narrative voice. Its disadvantages can include overlooking the emotions of characters, forgetting to include description, producing stilted writing, using too many sentence structures involving "I did . . .", and not devoting enough attention to the story's "here-and-now" immediacy.

Making Predictions and Inferences, and Drawing Conclusions

One technique authors often use to make their fictional stories more interesting is not giving away too much information easily; instead, they sprinkle in hints and descriptions. It is then up to the reader to draw a conclusion about the author's meaning by connecting textual clues with the reader's own pre-existing experiences and knowledge. **Drawing conclusions** is important as a reading strategy for understanding what is occurring in a text. Rather than directly stating who, what, where, when, or why, authors often describe story elements. Then, readers must draw conclusions to understand significant story components. As they go through a text, readers can think about the setting, characters, plot, problem, and solution; whether the author provided any clues for consideration; and combine any story clues with their existing knowledge and experiences to draw conclusions about what occurs in the text.

Making Predictions

Before and during reading, readers can apply the reading strategy of **making predictions** about what they think may happen next. For example, what plot and character developments will occur in fiction? What points will the author discuss in nonfiction? Making predictions about portions of text they have not yet read prepares readers mentally for reading, and also gives them a purpose for reading. To inform and make predictions about text, the reader can do the following:

- Consider the title of the text and what it implies
- Look at the cover of the book
- Look at any illustrations or diagrams for additional visual information
- Analyze the structure of the text
- Apply outside experience and knowledge to the text

Readers may adjust their predictions as they read. Reader predictions may or may not come true in the text, but as readers become more experienced as consumers of different texts, their ability to make accurate predictions will improve.

Making Inferences

Authors describe settings, characters, characters' emotions, and events. Readers must infer to understand a text fully. **Making inferences** enables readers to figure out meanings of unfamiliar words, make predictions about upcoming text, draw conclusions, and reflect on reading. Readers can infer about text before, during, and after reading. In everyday life, we use sensory information to infer. Readers can do the same with text. When authors do not answer all reader questions, readers must infer by saying "I think . . . This could be . . . This is because . . . Maybe . . . This means . . . I guess . . ." etc. Looking at illustrations, considering characters' behaviors, and asking questions during reading facilitate inference. Taking clues from text and connecting text to prior knowledge help to draw conclusions. Readers can infer word meanings, settings, reasons for occurrences, character emotions, pronoun referents, author messages, and answers to questions unstated in text. To practice making inferences, students can read sentences written/selected by the instructor, discuss the setting and character, draw conclusions, and make predictions.

Here is an example with three answer choices:

> Fred purchased the newest PC available on the market. Therefore, he purchased the most expensive PC in the computer store.
>
> What can one assume for this conclusion to follow logically?
>
> a. Fred enjoys purchasing expensive items.
> b. PCs are some of the most expensive personal technology products available.
> c. The newest PC is the most expensive one.

The premise of the text is the first sentence: *Fred purchased the newest PC.* The conclusion is the second sentence: *Fred purchased the most expensive PC.* Recent release and price are two different factors; the difference between them is the logical gap. To eliminate the gap, one must equate whatever new information the conclusion introduces with the pertinent information the premise has stated. This example simplifies the process by having only one of each: one must equate product recency with product price. Therefore, a possible bridge to the logical gap could be a sentence stating that the newest PCs always cost the most.

Making inferences and drawing conclusions involve skills that are quite similar: both require readers to fill in information the test writer has omitted. To make an inference or draw a conclusion about the text, test takers should observe all facts and arguments the test writer has presented. The best way to understand ways to drawing well-supported conclusions and generalizations is by practice. Let's take a look at the following example:

Nutritionist: More and more bodybuilders each year turn to whey protein as a source for their supplement intake to repair muscle tissue after working out. More and more studies are showing that using whey as a source of protein is linked to prostate cancer in men. Bodybuilders who use whey protein may consider switching to a plant-based protein source in order to avoid developing the negative effects that come with whey protein consumption.

Which of the following most accurately expresses the conclusion of the nutritionist's argument?

> a. Whey protein is an excellent way to repair muscles after a workout.
> b. Bodybuilders should switch from whey to a plant-based protein.
> c. Whey protein causes every single instance of prostate cancer in men.
> d. We still don't know the causes of prostate cancer in men.
> e. Whey protein is not effective at repairing muscle tissue after working out.

The correct answer choice is *B*: bodybuilders should switch from whey to a plant-based protein. We can gather this from the entirety of the passage, as it begins with what kind of protein bodybuilders consume, the dangers of that protein, and what kind of protein to switch to. Choice *A* is incorrect; this is the opposite of what the passage states. When reading through answer choices, it's important to look for choices that include the words "every," "always," or "all." In many instances, absolute answer choices will not be the correct answer. This example is shown in Choice *C*; the passage does not state that whey protein causes "every single instance" of prostate cancer in men, only that it is *linked* to prostate cancer in men. Choice *D* is incorrect; although the nutritionist doesn't list all the causes of prostate cancer in men, the nutritionist does not conclude that we don't know the causes of prostate cancer in men either. Finally, Choice *E* is incorrect because the argument does not provide any evidence that whey protein isn't effective at repairing muscle tissue, rather it raises concerns about the potentially dangerous side effect.

The key to drawing well-supported conclusions is to read the question stem in its entirety a few times over and then paraphrase the passage in your own words. Once you do this, you will get an idea of the passage's conclusion before you are confused by all the different answer choices. Remember that drawing a conclusion is different than making an assumption. With drawing a conclusion, we are relying solely on the passage for facts to come to our conclusion. Making an assumption goes beyond the facts of the passage, so be careful of answer choices depicting assumptions instead of passage-based conclusions.

Forming Hypotheses and Generalizations

Readers form **hypotheses** about the main idea as they read a passage. Hypotheses are based on incomplete evidence, so as the reader gathers more details from the passage, their hypothesis changes or is proven by the evidence. They start to form generalizations about the author's message.

Generalizations are broad conclusions based on the way the reader interprets the details. These generalizations might be correct, or the reader might have made inferences from the evidence that the author did not intend.

To determine if these generalizations were implied by the author, or if they are incorrect, the reader should examine the evidence they collected.

For example:

1. Humans cannot live without water.
2. People use thousands of gallons of water a year to keep their lawns green.

Based on these details, the reader concludes that *people who water their lawns are wasting water.*

The author has not offered evidence that watering lawns is wasteful, so the generalization is not supported. If the passage includes the detail, *scientists say water will soon be dangerously scarce*, the generalization is supported. However, the following statement would contradict the reader's generalization: *Lawns also need water to grow and sustain life.*

Understanding and Dissecting Arguments in Texts

Identifying the Author's Attitude or Viewpoint

The **author's attitude** toward a certain person or idea, or his or her purpose, may not always be stated. While it may seem impossible to know exactly what the author felt toward their subject, there are clues to indicate the emotion, or lack thereof, of the author. Clues like word choice or style will alert readers to the author's attitude. Some possible words that name the author's attitude are listed below:

- Admiring
- Angry
- Critical
- Defensive
- Enthusiastic
- Humorous
- Moralizing
- Neutral

- Objective
- Patriotic
- Persuasive
- Playful
- Sentimental
- Serious
- Supportive
- Sympathetic
- Unsupportive

An author's **tone** is the author's attitude toward their subject and is usually indicated by word choice. If an author's attitude toward their subject is one of disdain, the author will show the subject in a negative light, using deflating words or words that are negatively charged. If an author's attitude toward their subject is one of praise, the author will use agreeable words and show the subject in a positive light. If an author takes a neutral tone towards their subject, their words will be neutral as well, and they probably will show all sides of their subject, not just the negative or positive side.

Style is another indication of the author's attitude and includes aspects such as sentence structure, type of language, and formatting. Sentence structure is how a sentence is put together. Sometimes, short, choppy sentences will indicate a certain tone given the surrounding context, while longer sentences may serve to create a buffer to avoid being too harsh, or may be used to explain additional information. Style may also include formal or informal language. Using **formal language** to talk about a subject may indicate a level of respect. Using **informal language** may be used to create an atmosphere of friendliness or familiarity with a subject. Again, it depends on the surrounding context whether or not language is used in a negative or positive way. Style may also include formatting, such as determining the length of paragraphs or figuring out how to address the reader at the very beginning of the text.

The following is a passage from *The Florentine Painters of the Renaissance* by Bernhard Berenson. Following the passage is a question stem regarding the author's attitude toward their subject:

> Let us look now at an even greater triumph of movement than the Nudes, Pollaiuolo's "Hercules Strangling Antæus." As you realise the suction of Hercules' grip on the earth, the swelling of his calves with the pressure that falls on them, the violent throwing back of his chest, the stifling force of his embrace; as you realise the supreme effort of Antæus, with one hand crushing down upon the head and the other tearing at the arm of Hercules, you feel as if a fountain of energy had sprung up under your feet and were playing through your veins. I cannot refrain from mentioning still another masterpiece, this time not only of movement, but of tactile values and personal beauty as well—Pollaiuolo's "David" at Berlin. The young warrior has sped his stone, cut off the giant's head, and now he strides over it, his graceful, slender figure still vibrating with the rapidity of his triumph, expectant, as if fearing the ease of it. What lightness, what buoyancy we feel as we realise the movement of this wonderful youth!

Which one of the following best captures the author's attitude toward the paintings depicted in the passage?
- a. Neutrality towards the subject in this passage.
- b. Disdain for the violence found in the paintings.
- c. Excitement for the physical beauty found within the paintings.
- d. Passion for the movement and energy of the paintings.
- e. Disappointment in the quality of the paintings.

Choice *D* is the best answer. We know that the author feels positively about the subject because of the word choice. Berenson uses words and phrases like "supreme," "fountain of energy," "graceful," "figure still vibrating," "lightness," "buoyancy," and "wonderful youth." Notice also the exclamation mark at the end of the paragraph. These words and style depict an author full of passion, especially for the movement and energy found within the paintings.

Choice *A* is incorrect because the author is biased towards the subject due to the energy he writes with—he calls the movement in the paintings "wonderful" and by the other word choices and phrases, readers can tell that this is not an objective analysis of these paintings. For this reason, Choice *E* is clearly wrong; the author is impressed with the paintings, not disappointed. Choice *B* is incorrect because, although the author does mention the "violence" in the stance of Hercules, he does not exude disdain towards this. Choice *C* is incorrect. There is excitement in the author's tone, and some of this excitement is directed towards the paintings' physical beauty. However, this is not the *best* answer choice. Choice *D* is more accurate when stating the passion is for the movement and energy of the paintings, of which physical beauty is included.

Determining Whether Facts or Ideas are Relevant to an Argument

Evidence used in arguments must be credible and valid, such as that from peer-reviewed scholarly journals. Peer-reviewed sources are sources that have been reviewed by other experts in the field. Sources must also be relevant by pertaining to the argument posed and being up-to-date, especially those within the science or technology fields.

For example, let's consider a passage that discusses pesticides and the collapse of bees. An argument without relevant examples would look like this:

> With the use of the world's most popular pesticides, bees are becoming extinct. This is also causing ecological devastation. We must do something soon about the bee population, or else we will chase bees to extinction and lose valuable resources as a result.

Here is the same argument with examples. The added examples are in italics:

> With the use of the world's most popular pesticides, bees are becoming extinct. *Beekeepers have reported losing 55 to 95 percent of their colony in just two short years due to toxic poisoning.* This is also causing ecological devastation. *Bees are known for pollinating more than two-thirds of the world's most essential crops.* We must do something soon about the bee population, or else we will chase bees to extinction and lose valuable resources as a result.

Adding examples to the above argument brings life to the bees—they are living, dying, pollinating—and readers feel more compelled to act as a result of adding relevant examples to the argument.

When authors want to strengthen the support for an argument, **evidence-based data** in the form of statistics or concrete examples can be used. Statistics and examples are often accompanied by detailed explanations to help increase the audience's understanding and shape their ideas. Expert opinions are another way to strengthen an argument. But all this effort toward supporting a given argument does not necessarily make the argument absolute. After all, the word "argument" implies that there is more than one way to think about the subject. **Arguments** are meant to be challenged, questioned, and analyzed.

Authors will generally use one of two argument models: deductive or inductive. **Deductive arguments** require two general statements to support the argument. **Inductive arguments** employ specific data, examples, or facts to support the argument.

Deductive	Inductive
All fruits contain seeds. Tomatoes contain seeds. Therefore, tomatoes are fruits.	9 out of 10 dentists prefer soft-bristled toothbrushes. Therefore, soft-bristled toothbrushes are the best type of toothbrush for optimal dental health.

No matter what evidence is presented, readers should still challenge the argument. In any text, readers are encouraged to ask specific questions to evaluate the overall validity of the argument. Some important points to consider include:

- Has the author employed logic in the argument?
- Is the argument clearly explained?
- Is the argument sufficiently supported?
- Do the examples and statistics used actually pertain to the author's opinion or argument?
- Who conducted the research, and for what purpose?
- Is the supporting data qualitative, quantitative, or a mixture of both?
- Is the presented data representative of the typical cross-section of society or of the phenomenon being discussed?
- Does the author present any bias?
- Has the author overlooked anything that should be explored to form a well-rounded argument?

Although informational writing should be written objectively, such writing still constitutes the author's particular point of view or belief about a given subject. The author's main idea will likely be backed up with reasons, evidence, and supporting details, but it is important for the audience to question the main idea and evaluate the presented evidence. Although the author's ideas and shared details drive the overall argument, readers should feel compelled to explore the topic further, assess the evidence, and determine whether they agree with the overall message. Authors present the argument in order to convince their readers, but readers must strive to evaluate and assess the information to arrive at an informed opinion of the subject matter.

Persuasive Techniques Used By an Author

When authors write text for the purpose of persuading others to agree with them, they assume a position with the subject matter about which they are writing. Rather than presenting information objectively, the author treats the subject matter subjectively so that the information presented supports his or her position. In their argumentation, the author presents information that refutes or weakens opposing positions. Another technique authors use in persuasive writing is to anticipate arguments against the position. When students learn to read subjectively, they gain experience with the concept of persuasion in writing, and learn to identify positions taken by authors. This enhances their reading comprehension and develops their skills for identifying pro and con arguments and biases. Students should challenge the statements and opinions presented in a text by fully analyzing the argument or opinion stated and evaluating the evidence used by the author to support his or her stance.

There are five main parts of the **classical argument** that writers employ in a well-designed stance:

- Introduction: In the introduction to a classical argument, the author establishes goodwill and rapport with the reading audience, warms up the readers, and states the thesis or general theme of the argument.

- Narration: In the narration portion, the author gives a summary of pertinent background information, informs the readers of anything they need to know regarding the circumstances and environment surrounding and/or stimulating the argument, and establishes what is at risk or the stakes in the issue or topic. Literature reviews are common examples of narrations in academic writing.

- Confirmation: The confirmation states all claims supporting the thesis and furnishes evidence for each claim, arranging this material in logical order—e.g. from most obvious to most subtle or strongest to weakest.

- Refutation and Concession: The refutation and concession discuss opposing views and anticipate reader objections without weakening the thesis, yet permitting as many oppositions as possible.

- Summation: The summation strengthens the argument while summarizing it, supplying a strong conclusion and showing readers the superiority of the author's solution.

Introduction

A classical argument's introduction must pique reader interest, get readers to perceive the author as a writer, and establish the author's position. Shocking statistics, new ways of restating issues, or quotations or anecdotes focusing the text can pique reader interest. Personal statements, parallel instances, or analogies can also begin introductions—so can bold thesis statements if the author believes readers will agree. Word choice is also important for establishing author image with readers.

The introduction should typically narrow down to a clear, sound thesis statement. If readers cannot locate one sentence in the introduction explicitly stating the writer's position or the point they support, the writer probably has not refined the introduction sufficiently.

Narration and Confirmation

The narration part of a classical argument should create a context for the argument by explaining the issue to which the argument is responding, and by supplying any background information that influences the issue. Readers should understand the issues, alternatives, and stakes in the argument by the end of the narration to enable them to evaluate the author's claims equitably. The confirmation part of the classical argument enables the author to explain why they believe in the argument's thesis. The author builds a chain of reasoning by developing several individual supporting claims and explaining why that evidence supports each claim and also supports the overall thesis of the argument.

Refutation and Concession and Summation

The classical argument is the model for argumentative/persuasive writing, so authors often use it to establish, promote, and defend their positions. In the refutation aspect of the refutation and concession part of the argument, authors disarm reader opposition by anticipating and answering their possible objections, persuading them to accept the author's viewpoint. In the concession aspect, authors can concede those opposing viewpoints with which they agree. This can avoid weakening the author's thesis while establishing reader respect and goodwill for the author: all refutation and no concession can antagonize readers who disagree with the author's position. In the conclusion part of the classical argument, a less skilled writer might simply summarize or restate the thesis and related claims; however, this does not provide the argument with either momentum or closure. More skilled authors revisit the issues and the narration part of the argument, reminding readers of what is at stake.

Distinguishing Between Facts and Opinions

A **fact** is a statement that is true empirically or an event that has actually occurred in reality, and can be proven or supported by evidence; it is generally objective. In contrast, an **opinion** is subjective, representing something that someone believes rather than something that exists in the absolute. People's individual understandings, feelings, and perspectives contribute to variations in opinion. Though facts are typically objective in nature, in some instances, a statement of fact may be both factual and yet also subjective. For example, emotions are individual subjective experiences. If an individual says that he or she feels happy or sad, the feeling is subjective, but the statement is factual; hence, it is a subjective fact. In contrast, if one person tells another that the other is feeling happy or sad—whether this is true or not—it is an assumption or an opinion.

Authors use arguments to persuade the reader to agree with their claim. When analyzing the strength of an argument, the reader should summarize the author's message and determine whether or not they agree or disagree with the claim and whether, overall, they believe the evidence the author presented to support it. Next, the reader should identify each individual detail and examine it to determine if it supports the claim, fails to support the claim, or even distract from the author's goal.

As readers examine each piece of evidence, they should consider its type and its effectiveness in proving the author's claim both individually and in context. Readers should pay attention to details that contradict the evidence, and readers should question the way the author uses details.

Facts can often be interpreted in ways that mislead the reader, so they will believe a claim that is unproven or untrue. If the claim is supported by facts such as statistics or empirical evidence, the reader should examine them. The reader should question whether the facts are relevant or, if they were interpreted differently, they would lead to a different conclusion that would fail to support the argument or even contradict it. They should consider whether the scope of the data is appropriate, or if it is too broad or narrow, and they should question whether the source is credible, and whether a source has been cited at all.

If the author uses expert opinions, the reader should consider whether the expert is credible and appropriate. If a doctor offers an opinion about pediatric medicine, the reader should ask if the doctor is an expert in that field or if they practice a different form of medicine.

Finally, the reader should consider whether the author has anticipated and adequately responded to any potential counterarguments to the claim and decide if the argument overall is strong, or if the evidence fails to persuade them to accept the author's claim.

Challenging the Statements and Opinions Presented in a Text

Selecting the most relevant material to support a written text is a necessity in producing quality writing and for the credibility of an author. Arguments lacking in reasons or examples won't work in persuading the audience later on, because their hearts have not been pulled. Using examples to support ideas also gives the writing rhetorical effects such as **pathos** (emotion), **logos** (logic), or **ethos** (credibility), all three of which are necessary for a successful text.

Not all arguments are valid. Authors sometimes have one or more flaws in their argument's reasoning. Critical readers should always consider the validity and logic underlying the statements and opinions in a text. For example, texts wrought with bias or stereotypes should be read with caution and evaluated critically, rather than read complacently.

Biases usually occur when someone allows their personal preferences or ideologies to interfere with what should be an objective decision. In personal situations, someone is biased towards someone if they favor them in an unfair way. In academic writing, being biased in your sources means leaving out objective information that would turn the argument one way or the other. The evidence of bias in academic writing makes the text less credible, so be sure to present all viewpoints when writing, not just your own, so to avoid coming off as biased. Being objective when presenting information or dealing with people usually allows the person to gain more credibility.

Stereotypes are preconceived notions that place a particular rule or characteristics on an entire group of people. Stereotypes are usually offensive to the group they refer to or allies of that group, and often have negative connotations. The reinforcement of stereotypes isn't always obvious. Sometimes stereotypes can be very subtle and are still widely used in order for people to understand categories within the world. For example, saying that women are more emotional and intuitive than men is a stereotype, although this is still an assumption used by many in order to understand the differences between one another.

In addition, critical readers are able to identify flaws or holes in an argument's reasoning. Some questions may ask you to provide a description of that error. In order for you to be able to describe what flaw is occurring in the argument, it will help to know of various argumentative flaws, such as red herring, false choice, and correlation vs. causation.

Here are some examples of what this type of question looks like:

- The reasoning in the author's argument is flawed because the argument . . .
- The argument is most vulnerable to criticism on the grounds that it . . .
- Which one of the following is an error in the argument's reasoning?
- A flaw in the reasoning of the argument is that . . .
- Which one of the following most accurately describes X's criticism of the argument made by Y?

Bait/Switch

One common flaw that is good to know is called the **bait and switch**. It occurs when the author will provide an argument that offers evidence about X, and ends the argument with a conclusion about Y. A "bait and switch" answer choice will look like this:

The argument assumes that X does in fact address Y without providing justification.

Let's look at an example:

Hannah will most likely always work out and maintain a healthy physique. After all, Hannah's IQ is extremely high.

The correct answer will look like this:

The argument assumes that Hannah's high IQ addresses her likelihood of always working out without providing justification.

Ascriptive Error

The **ascriptive argument** will begin the argument with something a third party has claimed. Usually, it will be something very general, like "Some people say that . . ." or "Generally, it has been said . . ." Then, the arguer will follow up that claim with a refutation or opposing view. The problem here is that when the arguer phrases something in this general sense without a credible source, their refutation of that evidence doesn't really matter. Here's an example:

> It has been said that peppermint oil has been proven to relieve stomach issues and, in some cases, prevent cancer. I can attest to the relief in stomach issues; however, there is just not enough evidence to prove whether or not peppermint oil has the ability to prevent any kind of cancerous cells from forming in the body.

The correct answer will look like this:

> The argument assumes that the refuting evidence matters to the position that is being challenged.

> We have no credible source in this argument, so the refutation is senseless.

Prescriptive Error

First, let's take a look at what "prescriptive" means. **Prescriptive** argument means to give directions, or to say something *ought to* or *should* do something else. Sometimes an argument will be a descriptive premise (simply describing) that leads to a prescriptive conclusion, which makes for a very weak argument. This is like saying "There is a hurricane coming; therefore, we should leave the state." Even though this seems like common sense, the logical soundness of this argument is missing. A valid argument is when the truth of the premise leads absolutely to the truth of the conclusion. It's when the conclusion *is* something, not when the conclusion *should* be or do something. The flaw here is the assumption that the conclusion is going to work out; something prescriptive is not ever guaranteed to work out in a logical argument.

False Choice

A **false choice flaw**, or false dilemma flaw, is a statement that assumes only the object it lists in the statement is the solution, or the only options that exist, for that problem. Here is an example:

> I didn't get the grade I want in Chemistry class. I must either be really stupid, I didn't get enough sleep, or I didn't eat enough that day.

This is a false choice error. We are offered only three options for why the speaker did not get the grade he or she wanted in Chemistry class. However, there is potentially more options why the grade was not achieved other than the three listed. The speaker could have been fighting a cold, or the professor may not have taught the material in a comprehensive way. It is our job as test takers to recognize that there are more options other than the choices we are given, although it appears that the only three choices are listed in the example.

Red Herring

A **red herring** is a point offered in an argument that is only meant to distract or mislead. A red herring will throw something out after the argument that is unrelated to the argument, although it still commands attention, thus taking attention away from the relevant issue. The following is an example of a red herring fallacy:

> Kirby: It seems like therapy is moving toward a more holistic model rather than something prescriptive, where the space between a therapist and client is seen more organic rather than a controlled space. This helps empower the client to reach their own conclusions about what should be done rather than having someone tell them what to do.

> Barlock: What's the point of therapy anyway? It seems like "talking out" problems with a stranger is a waste of time and always has been. Is it even successful as a profession?

We see Kirby present an argument about the route therapy is taking toward the future. Instead of responding to the argument by presenting their own side regarding where therapy is headed, Barlock questions the overall point of therapy. Barlock throws out a red herring here: Kirby cannot proceed with the argument because now Kirby must defend the existence of therapy instead of its future.

Correlation Versus Causality

Readers should be careful when reviewing causal conclusions because the reasoning is often flawed, incorrectly classifying **correlation as causality**. Two events that may or may not be associated with one another are said to be linked such that one was the cause or reason for the other, which is considered the effect. To be a true "cause-and-effect" relationship, one factor or event must occur first (the cause) and be the sole reason (unless others are also listed) that the other occurred (the effect). The cause serves as the initiator of the relationship between the two events.

For example, consider the following argument:

> Last weekend, the local bakery ran out of blueberry muffins and some customers had to select something else instead. This week, the bakery's sales have fallen. Therefore, the blueberry muffin shortage last weekend resulted in fewer sales this week.

In this argument, the author states that the decline in sales this week (the effect) was caused by the shortage of blueberry muffins last weekend. However, there are other viable alternate causes for the decline in sales this week besides the blueberry muffin shortage. Perhaps it is summer, and many normal patrons are away this week on vacation, or maybe another local bakery just opened or is running a special sale this week. There might be a large construction project or road work in town near the bakery, deterring customers from navigating the detours or busy roads. It is entirely possible that the decline this week is just a random coincidence and not attributable to any factor other than chance, and that next week, sales will return to normal or even exceed typical sales. Insufficient evidence exists to confidently assert that the blueberry muffin shortage was the sole reason for the decline in sales, thus mistaking correlation for causation.

Vocabulary

Determining the Meaning of Unknown Words

Context Clues

When readers encounter an unfamiliar word in text, they can use the surrounding **context**—the overall subject matter, specific chapter/section topic, and especially the immediate sentence context. Among others, one category of **context clues** is grammar. For example, the position of a word in a sentence and its relationship to the other words can help the reader establish whether the unfamiliar word is a verb, a noun, an adjective, an adverb, etc. This narrows down the possible meanings of the word to one part of speech. However, this may be insufficient. In a sentence that many birds *migrate* twice yearly, the reader can determine the word is a verb, and probably does not mean eat or drink; but it could mean travel, mate, lay eggs, hatch, molt, etc.

Another common context clue is a synonym or definition included in the sentence. Sometimes both exist in the same sentence. Here's an example:

Scientists who study birds are *ornithologists*.

Many readers may not know the word *ornithologist*. However, the example contains a definition (scientists who study birds). The reader may also have the ability to analyze the suffix (-*logy*, meaning the study of) and root (*ornitho-*, meaning bird).

Another common context clue is a sentence that shows differences. Here's an example:

Birds *incubate* their eggs outside of their bodies, unlike mammals.

Some readers may be unfamiliar with the word *incubate*. However, since we know that "unlike mammals," birds incubate their eggs outside of their bodies, we can infer that *incubate* has something to do with keeping eggs warm outside the body until they are hatched.

In addition to analyzing the etymology of a word's root and affixes and extrapolating word meaning from sentences that contrast an unknown word with an antonym, readers can also determine word meanings from sentence context clues based on logic. Here's an example:

Birds are always looking out for predators that could attack their young.

The reader who is unfamiliar with the word *predator* could determine from the context of the sentence that predators usually prey upon baby birds and possibly other young animals. Readers might also use the context clue of etymology here, as *predator* and *prey* have the same root.

Figurative and Colloquial Language

Authors of a text use language with multiple levels of meaning for many different reasons. When the meaning of a text calls for directness, **literal language** should be used to provide clarity to the reader. **Figurative language** can be used when the author wants to produce an emotional effect in the reader or facilitate a deeper understanding of a word or passage. For example, if someone wanted to write a set of instructions on how to use a computer, they would write in literal language. However, if someone wanted to comment on the social implications of banning immigration, they might want to use a wide range of figurative language to highlight an empathetic response. It is important to keep in mind, too, that a single text can have a mixture of both literal and figurative language.

It's important to be able to recognize and interpret figurative, or non-literal, language. Literal statements rely directly on the denotations of words and express exactly what's happening in reality. Figurative language uses nonliteral expressions to present information in a creative way. Consider the following sentences:

> a. His pillow was very soft, and he fell asleep quickly.

> b. His pillow was a fluffy cloud, and he floated away on it to the dream world.

Sentence *A* is literal, employing only the real meanings of each word. Sentence *B* is figurative. It employs a metaphor by stating that his pillow was a cloud. Of course, he isn't actually sleeping on a cloud, but the reader can draw on images of clouds as light, soft, fluffy, and relaxing to get a sense of how the character felt as he fell asleep. Also, in sentence *B*, the pillow becomes a vehicle that transports him to a magical dream world. The character isn't literally floating through the air—he's simply falling asleep! But by utilizing figurative language, the author creates a scene of peace, comfort, and relaxation that conveys stronger emotions and more creative imagery than the purely literal sentence.

Figurative language is used more heavily in texts such as literary fiction, poetry, critical theory, and speeches. Figurative language goes beyond literal language, allowing readers to form associations they wouldn't normally form with literal language. Using language in a figurative sense appeals to the imagination of the reader. It is important to remember that words themselves are signifiers of objects and ideas, and not the objects and ideas themselves. Figurative language can highlight this detachment by creating multiple associations, but also points to the fact that language is fluid and capable of creating a world full of linguistic possibilities. Figurative language, it can be argued, is the heart of communication even outside of fiction and poetry. People connect through humor, metaphors, cultural allusions, puns, and symbolism in their everyday rhetoric. The following are terms associated with figurative language:

A **simile** is a comparison of two things using *like*, *than*, or *as*. A simile usually takes objects that have no apparent connection, such as a mind and an orchid, and compares them:

> His mind was as complex and rare as a field of ghost orchids.

Similes encourage a new, fresh perspective on objects or ideas that wouldn't otherwise occur. Similes are different than metaphors. Metaphors do not use *like*, *than*, or *as*. So, a metaphor from the above example would be:

> His mind was a field of ghost orchids.

Thus, similes highlight the comparison by focusing on the figurative side of the language, elucidating more the author's intent: a field of ghost orchids is something complex and rare, like the mind of a genius. With the metaphor, however, we get a beautiful yet somewhat equivocal comparison.

A popular use of figurative language, a **metaphor** compares objects or ideas directly, asserting that something *is* a certain thing, even if it isn't. The following is an example of a metaphor used by writer Virginia Woolf:

> Books are the mirrors of the soul.

Metaphors have a vehicle and a tenor. The tenor is "books" and the vehicle is "mirrors of the soul." That is, the tenor is what is meant to be described, and the vehicle is that which carries the weight of the comparison. In this metaphor, perhaps the author means to say that written language (books) reflect a person's most inner thoughts and desires.

There are also **dead metaphors**, which means that the phrases have been so overused to the point where the figurative meaning becomes literal, like the phrase "What you're saying is crystal clear." The phrase compares "what's being said" to something "crystal clear." However, since the latter part of the phrase is in such popular use, the meaning seems literal ("I understand what you're saying") even when it's not.

Finally, an **extended metaphor** is a metaphor that goes on for several paragraphs, or even an entire text. John Keats' poem "On First Looking into Chapman's Homer" begins, "Much have I travell'd in the realms of gold," and goes on to explain the first time he hears Chapman's translation of Homer's writing. We see the extended metaphor begin in the first line. Keats is comparing travelling into "realms of gold" and exploration of new lands to the act of hearing a certain kind of literature for the first time. The extended metaphor goes on until the end of the poem where Keats stands "Silent, upon a peak in Darien," having heard the end of Chapman's translation. Keats has gained insight into new lands (new text) and is the richer for it.

The following are brief definitions and examples of popular figurative language:

- **Onomatopoeia**: A word that, when spoken, imitates the sound to which it refers. Example: "We heard a loud *boom* while driving to the beach yesterday."

- **Personification**: When human characteristics are given to animals, inanimate objects, or abstractions. An example would be in William Wordsworth's poem "Daffodils" where he sees a "crowd . . . / of golden daffodils . . . / Fluttering and dancing in the breeze." Dancing is usually a characteristic attributed solely to humans, but Wordsworth personifies the daffodils here as a crowd of people dancing.

- **Juxtaposition**: Juxtaposition is placing two objects side by side for comparison. In literature, this might look like placing two characters side by side for contrasting effect, like God and Satan in Milton's "Paradise Lost."

- **Paradox**: A paradox is a statement that is self-contradictory but will be found nonetheless true. One example of a paradoxical phrase is when Socrates said "I know one thing; that I know nothing." Seemingly, if Socrates knew nothing, he wouldn't know that he knew nothing. However, it is one thing he knows: that true wisdom begins with casting all presuppositions one has about the world aside.

- **Hyperbole**: A hyperbole is an exaggeration. Example: "I'm so tired I could sleep for centuries."

- **Allusion**: An allusion is a reference to a character or event that happened in the past. An example of a poem littered with allusions is T.S. Eliot's "The Waste Land." An example of a biblical allusion manifests when the poet says, "I will show you fear in a handful of dust," creating an ominous tone from Genesis 3:19 "For you are dust, and to dust you shall return."

- **Pun**: Puns are used in popular culture to invoke humor by exploiting the meanings of words. They can also be used in literature to give hints of meaning in unexpected places. One example of a pun is when Mercutio is giving his monologue after he is stabbed by Tybalt in "Romeo and Juliet" and says, "look for me tomorrow and you will find me a grave man."

- **Imagery**: This is a collection of images given to the reader by the author. If a text is rich in imagery, it is easier for the reader to imagine themselves in the author's world.

One example of a poem that relies on *imagery* is William Carlos Williams' "The Red Wheelbarrow":

> so much depends
> upon
>
> a red wheel
> barrow
>
> glazed with rain
> water
>
> beside the white
> chickens

The starkness of the imagery and the placement of the words in the poem, to some readers, throw the poem into a meditative state where, indeed, the world of this poem is made up solely of images of a purely simple life. This poem tells a story in sixteen words by using imagery.

- **Symbolism**: A symbol is used to represent an idea or belief system. For example, poets in Western civilization have been using the symbol of a rose for hundreds of years to represent love. In Japan, poets have used the firefly to symbolize passionate love, and sometimes even spirits of those who have died. Symbols can also express powerful political commentary and can be used in propaganda.

- **Irony**: There are three types of irony. Verbal irony is when a person states one thing and means the opposite. For example, a person is probably using irony when they say, "I can't wait to study for this exam next week." Dramatic irony occurs in a narrative and happens when the audience knows something that the characters do not. In the modern TV series *Hannibal*, we as an audience know that Hannibal Lecter is a serial killer, but most of the main characters do not. This is dramatic irony. Finally, situational irony is when one expects something to happen, and the opposite occurs. For example, we can say that a fire station burning down would be an instance of situational irony.

Lastly, **idiomatic expressions** are phrases or groups of words that have an established meaning when used together that's unrelated to the literal meanings of the individual words. For example, consider the following sentence that includes a common idiomatic phrase:

> I know Phil is coming to visit this weekend because I heard it straight from the horse's mouth.

The speaker of this sentence did not consult a horse nor hear anything uttered from a horse in relation to Phil's visit. Instead, "straight from the horse's mouth" is an **idiom** that means it's the truth. As in the sentence in which it is used above, it often means whatever said should be taken as truth because it was spoken by a reliable source or by the person to which it pertains (in this case, Phil). The phrase is derived from the fact that sellers of horses at auctions would sometimes try to lie about the age of the horse. However, the size and shape of a horse's teeth can provide a fairly accurate estimate of the horse's true age. Therefore, essentially the truth regarding the horse's age come straight from their mouth. The idiomatic expression came to mean getting the truth in any situation.

Finding errors in idiomatic expressions can be difficult because it requires familiarity with the idiom. Because there are more than one thousand idioms in the English language, memorizing all of them is impractical. However, it is recommended to review the most common ones. There are many webpages dedicated to listing and explaining frequently used idioms.

Errors in the idiomatic expressions are typically one of two types. Either the idiomatic expression is stated improperly or it is used incorrectly. In the first type of issue, the prepositions used are often incorrect. Using the idiom from above, for example, it might say "straight *in* the horse's mouth" or "straight *with* the horse's mouth." In the second error type, the idiomatic expression is used improperly because the meaning it carries does not make sense in the context in which it is being used.

Consider the following:

> He was looking straight from the horse's mouth when he complained about the phone his father bought him.

Here, the writer has confused the idiom "straight from the horse's mouth" with "looking a gift horse in the mouth," which means to find fault in a gift or favor.

Either type of error can be difficult to detect and correct without prior knowledge of the idiomatic phrase. Practicing using idioms and discussing and studying their origins can help students remember the meanings and precise wordings, which will help them identify and correct errors in their usage.

Different Interpretations That Can Be Made of the Same Word or Phrase

Language can function differently in different contexts. The same words can convey meaning in nuanced ways depending on the surrounding context and the style and tone of the composition. Just as how people can speak with a variety of tones and inflections that can alter the meaning of the same sentence, so too can writers insert tone into written words. Punctuation choice is one example of how the same sentence can be interpreted slightly differently.

Consider the following three sentences:

> Camille hates dogs.

> Camille hates dogs!

> Camille hates dogs?

Although the wording is identical in these three simple sentences, the end punctuation choice provides slightly different interpretations because the tone is affected. The first is a statement. It lacks a significant overbearing tone. It's simply stating that Camille hates dogs, but readers aren't guided as to an emotional response to that fact. The second sentence ends in an exclamation point. This punctuation tends to evoke a feeling or surprise, exasperation, or urgency and alarm. Perhaps, for example, someone was about to introduce a large mastiff to Camille who had her back turned to the dog. One of Camille's friends who saw what was about to happen may have shouted that sentence in caution to prevent a terrified Camille. The last sentence is obviously a question, but that doesn't mean it's affectless. It may be said in an incredulous tone or one of surprise. The speaker may be a dog lover and find it hard to believe Camille is not. It might be uttered with doubt.

The writer could further display tone by italicizing one of the words to indicate emphasis. Taking just the question as an example, consider the difference between the following three examples:

Camille hates dogs?

Camille *hates* dogs?

Camille hates *dogs*?

The first example places the emphasis on Camille. The speaker could be surprised or trying to clarify that it is Camille, not another person, who supposedly hates dogs. In the second example, the speaker's focus is on the word *hate*. He or she is seeking clarity or confirmation that Camille actually hates dogs (rather than loves them, likes them, is afraid of them, etc.). The italicized *dogs* in the last sentence indicates the speaker is verifying or shocked that it is dogs, in particular, that Camille hates (rather than cats, spiders, rats, etc.). Other punctuation marks, especially commas, can also shape the way a sentence is read and interpreted.

In addition to punctuation and emphasis indicators, the same words can be interpreted differently based on their context. **Denotation** refers to a word's explicit definition, like that found in the dictionary. Denotation is often set in comparison to connotation. **Connotation** is the emotional, cultural, social, or personal implication associated with a word. Denotation is more of an objective definition, whereas connotation can be more subjective, although many connotative meanings of words are similar for certain cultures. The denotative meanings of words are usually based on facts, and the connotative meanings of words are usually based on emotion. Here are some examples of words and their denotative and connotative meanings in Western culture:

Word	Denotative Meaning	Connotative Meaning
Home	A permanent place where one lives, usually as a member of a family.	A place of warmth; a place of familiarity; comforting; a place of safety and security. "Home" usually has a positive connotation.
Snake	A long reptile with no limbs and strong jaws that moves along the ground; some snakes have a poisonous bite.	An evil omen; a slithery creature (human or nonhuman) that is deceitful or unwelcome. "Snake" usually has a negative connotation.
Winter	A season of the year that is the coldest, usually from December to February in the northern hemisphere and from June to August in the southern hemisphere.	Circle of life, especially that of death and dying; cold or icy; dark and gloomy; hibernation, sleep, or rest. Winter can have a negative connotation, although many who have access to heat may enjoy the snowy season from their homes.

For one final example, consider how the word *fly* has a different meaning in each of the following sentences:

- "His trousers have a fly on them."
- "He swatted the fly on his trousers."
- "Those are some fly trousers."
- "They went fly fishing."
- "She hates to fly."
- "If humans were meant to fly, they would have wings."

As strategies, readers can try substituting a familiar word for an unfamiliar one and see whether it makes sense in the sentence. They can also identify other words in a sentence, offering clues to an unfamiliar word's meaning.

Practice Questions

The next two questions are based off the following passage:

Rehabilitation rather than punitive justice is becoming much more popular in prisons around the world. Prisons in America, especially, where the recidivism rate is 67 percent, would benefit from mimicking prison tactics in Norway, which has a recidivism rate of only 20 percent. In Norway, the idea is that a rehabilitated prisoner is much less likely to offend than one harshly punished. Rehabilitation includes proper treatment for substance abuse, psychotherapy, healthcare and dental care, and education programs.

1. The author's purpose is _____
   a. to show the audience one of the effects of criminal rehabilitation by comparison
   b. to persuade the audience to donate to American prisons for education programs
   c. to convince the audience of the harsh conditions of American prisons
   d. to inform the audience of the incredibly lax system of Norway prisons

2. The word *recidivism*, as used in the passage, means _____
   a. the lack of violence in the prison system.
   b. the opportunity of inmates to receive therapy in prison.
   c. the event of a prisoner escaping the compound.
   d. the likelihood of a convicted criminal to reoffend.

The next three questions are based off the following passage from Virginia Woolf's Mrs. Dalloway:

What a lark! What a plunge! For so it had always seemed to her, when, with a little squeak of the hinges, which she could hear now, she had burst open the French windows and plunged at Bourton into the open air. How fresh, how calm, stiller than this of course, the air was in the early morning; like the flap of a wave; the kiss of a wave; chill and sharp and yet (for a girl of eighteen as she then was) solemn, feeling as she did, standing there at the open window, that something awful was about to happen; looking at the flowers, at the trees with the smoke winding off them and the rooks rising, falling; standing and looking until Peter Walsh said, "Musing among the vegetables?"— was that it? —"I prefer men to cauliflowers"— was that it? He must have said it at breakfast one morning when she had gone out on to the terrace — Peter Walsh. He would be back from India one of these days, June or July, she forgot which, for his letters were awfully dull; it was his sayings one remembered; his eyes, his pocket-knife, his smile, his grumpiness and, when millions of things had utterly vanished — how strange it was! — a few sayings like this about cabbages.

3. The passage above is _____ writing.
   a. persuasive
   b. expository
   c. narrative
   d. technical

4. The narrator was feeling _____ right before Peter Walsh's voice distracted her.
    a. a spark of excitement for the morning
    b. anger at the larks
    c. a sense of foreboding
    d. confusion at the weather

5. The main idea of the passage is _____
    a. to present the events leading up to a party.
    b. to show the audience that the narrator is resentful towards Peter.
    c. to introduce Peter Walsh back into the narrator's memory.
    d. to reveal what mornings are like in the narrator's life.

6. Which of the following is an acceptable heading to insert into the blank space?

Chapter 5: Literature and Language Usage
    I. Figurative Speech
            1. Alliteration
            2. Metaphors
            3. Onomatopoeia
            4. _____
    a. Nouns
    b. Agreement
    c. Personification
    d. Syntax

7. Felicia knew she had to be <u>prudent</u> if she was going to cross the bridge over the choppy water; one wrong move and she would be falling toward the rocky rapids.

In the context above, prudent most nearly means _____

    a. patient.
    b. afraid.
    c. dangerous.
    d. careful.

The next three questions are based on the passage from Many Marriages *by Sherwood Anderson.*

There was a man named Webster who lived in a town of twenty-five thousand people in the state of Wisconsin. He had a wife named Mary and a daughter named Jane and he was himself a fairly prosperous manufacturer of washing machines. When the thing happened of which I am about to write, he was thirty-seven or thirty-eight years old and his one child, the daughter, was seventeen. Of the details of his life up to the time a certain revolution happened within him it will be unnecessary to speak. He was however a rather quiet man inclined to have dreams which he tried to crush out of himself in order that he function as a washing machine manufacturer; and no doubt, at odd moments, when he was on a train going some place or perhaps on Sunday afternoons in the summer when he went alone to the deserted office of the factory and sat several hours looking out at a window and along a railroad track, he gave way to dreams.

8. What does the author mean by the following sentence?

"Of the details of his life up to the time a certain revolution happened within him it will be unnecessary to speak."

a. The details of his external life don't matter; only the details of his internal life matter.
b. Whatever happened in his life before he had a certain internal change is irrelevant.
c. He had a traumatic experience earlier in his life which rendered it impossible for him to speak.
d. Before the revolution, he was a lighthearted man who always wished to speak to others no matter who they were.

9. The point of view this narrative is told in is _____
    a. first person limited.
    b. first person omniscient.
    c. second person.
    d. third person.

10. Webster is a _____
    a. washing machine manufacturer.
    b. train operator.
    c. leader of the revolution.
    d. stay-at-home husband.

11. After Sheila recently had a coronary artery bypass, her doctor encouraged her to switch to a plant-based diet to avoid foods loaded with cholesterol and saturated fats. Sheila's doctor has given her a list of foods she can purchase in order to begin making healthy dinners, which excludes dairy (cheese, yogurt, cream) eggs, and meat. The doctor's list includes the following: pasta, marinara sauce, tofu, rice, black beans, tortilla chips, guacamole, corn, salsa, rice noodles, stir-fry vegetables, teriyaki sauce, quinoa, potatoes, yams, bananas, eggplant, pizza crust, cashew cheese, almond milk, bell pepper, and tempeh.

Sheila can make _____ and still follow the doctor's orders.

    a. eggplant parmesan with a salad
    b. veggie pasta with marinara sauce
    c. egg omelet with no cheese and bell peppers
    d. quinoa burger with cheese and French fries

12. Ecologist: If we do not act now, more than one hundred animal species will be extinct by the end of the decade. The best way to save them is to sell hunting licenses for endangered species. Hunters can pay for the right to kill old and lame animals. Otherwise, there's no way to fund our conservation efforts.

Which one of the following assumptions does the ecologist's argument rely upon?
    a. Hunting licenses for non-endangered species aren't profitable.
    b. All one hundred animal species must be saved.
    c. The new hunting license revenue will fund conservation efforts.
    d. Conservation efforts should have begun last decade.

The following three questions are based on the book On the Trail *by Lina Beard and Adelia Belle Beard.*

> For any journey, by rail or by boat, one has a general idea of the direction to be taken, the character of the land or water to be crossed, and of what one will find at the end. So it should be in striking the trail. Learn all you can about the path you are to follow. Whether it is plain or obscure, wet or dry; where it leads; and its length, measured more by time than by actual miles. A smooth, even trail of five miles will not consume the time and strength that must be expended upon a trail of half that length which leads over uneven ground, varied by bogs and obstructed by rocks and fallen trees, or a trail that is all up-hill climbing. If you are a novice and accustomed to walking only over smooth and level ground, you must allow more time for covering the distance than an experienced person would require and must count upon the expenditure of more strength, because your feet are not trained to the wilderness paths with their pitfalls and traps for the unwary, and every nerve and muscle will be strained to secure a safe foothold amid the tangled roots, on the slippery, moss-covered logs, over precipitous rocks that lie in your path. It will take time to pick your way over boggy places where the water oozes up through the thin, loamy soil as through a sponge; and experience alone will teach you which hummock of grass or moss will make a safe stepping-place and will not sink beneath your weight and soak your feet with hidden water. Do not scorn to learn all you can about the trail you are to take . . . It is not that you hesitate to encounter difficulties, but that you may prepare for them. In unknown regions take a responsible guide with you, unless the trail is short, easily followed, and a frequented one. Do not go alone through lonely places; and, being on the trail, keep it and try no explorations of your own, at least not until you are quite familiar with the country and the ways of the wild.

13. The author says _____ about unknown regions.
    a. You should try and explore unknown regions in order to learn the land better.
    b. Unless the trail is short or frequented, you should take a responsible guide with you.
    c. All unknown regions will contain pitfalls, traps, and boggy places.
    d. It's better to travel unknown regions by rail rather than by foot.

14. _____ is NOT a detail from the passage.
    a. Learning about the trail beforehand is imperative
    b. Time will differ depending on the land
    c. Once you are familiar with the outdoors you can go places on your own
    d. Be careful for wild animals on the trail you are on

15. The passage above is considered _____ writing.
    a. descriptive
    b. persuasive
    c. narrative
    d. informative

16. Ethicist: Artificial intelligence is rapidly approaching consciousness. What began as simple algorithms that locate and regurgitate information is now capable of independently drawing conclusions. Unfortunately, the free market is responsible for the advances in artificial intelligence, and private companies aren't incentivized to align artificial intelligence with humanity's goals. Without any guidance, artificial intelligence will adopt humanity's worst impulses and mirror the Internet's most violent worldviews.

The ethicist would most likely agree that _____
   a. the Internet is negatively impacting society.
   b. unregulated artificial intelligence is a threat to humanity.
   c. humanity's goals should always be prioritized over technological advancements.
   d. artificial intelligence should be limited to simple algorithms.

17. Businessman: My cardinal rule is to only invest in privately-held small businesses that exclusively sell tangible goods and have no debt.

Which one of the following is the best investment opportunity according to the businessman's cardinal rule?
   a. Jose owns his own grocery store. He's looking for a partner, because he fell behind on his mortgage and owes the bank three months' worth of payments.
   b. Elizabeth is seeking a partner with business expertise to help expand her standalone store that sells niche board games. The store isn't currently profitable, but it's never been in debt.
   c. A family-owned accounting firm with no outstanding debts is looking for its first outside investor. The firm has turned a profit every year since it opened
   d. A multinational corporation is selling high-yield bonds for the first time.

The next seven questions are based on the following passage from The Story of Germ Life *by Herbert William Conn:*

The first and most universal change effected in milk is its souring. So universal is this phenomenon that it is generally regarded as an inevitable change which can not be avoided, and, as already pointed out, has in the past been regarded as a normal property of milk. To-day, however, the phenomenon is well understood. It is due to the action of certain of the milk bacteria upon the milk sugar which converts it into lactic acid, and this acid gives the sour taste and curdles the milk. After this acid is produced in small quantity its presence proves deleterious to the growth of the bacteria, and further bacterial growth is checked. After souring, therefore, the milk for some time does not ordinarily undergo any further changes.

Milk souring has been commonly regarded as a single phenomenon, alike in all cases. When it was first studied by bacteriologists it was thought to be due in all cases to a single species of micro-organism which was discovered to be commonly present and named *Bacillus acidi lactici.* This bacterium has certainly the power of souring milk rapidly, and is found to be very common in dairies in Europe. As soon as bacteriologists turned their attention more closely to the subject it was found that the spontaneous souring of milk was not always caused by the same species of bacterium. Instead of finding this *Bacillus acidi lactici* always present, they found that quite a number of different species of bacteria have the power of souring milk, and are found in different specimens of soured milk. The number of species of bacteria which have been found to sour milk has increased until something over a hundred are known to have this power. These different species do not affect the milk in the same way. All produce some acid, but they differ in the kind and the amount of acid, and especially in the other changes which are effected at the same time that the milk is soured, so that the resulting soured milk is quite variable. In spite of this variety, however, the most recent work tends to show that the majority of cases of spontaneous souring of milk are produced by bacteria which, though somewhat variable, probably constitute a single species, and are identical with the *Bacillus acidi lactici*. This species, found common in the dairies of Europe, according to recent investigations occurs in this country as well. We may say, then, that while there are many species of bacteria infesting the dairy which can sour the milk, there is one which is more common and more universally found than others, and this is the ordinary cause of milk souring.

When we study more carefully the effect upon the milk of the different species of bacteria found in the dairy, we find that there is a great variety of changes which they produce when they are allowed to grow in milk. The dairyman experiences many troubles with his milk. It sometimes curdles without becoming acid. Sometimes it becomes bitter, or acquires an unpleasant "tainted" taste, or, again, a "soapy" taste. Occasionally a dairyman finds his milk becoming slimy, instead of souring and curdling in the normal fashion. At such times, after a number of hours, the milk becomes so slimy that it can be drawn into long threads. Such an infection proves very troublesome, for many a time it persists in spite of all attempts made to remedy it. Again, in other cases the milk will turn blue, acquiring about the time it becomes sour a beautiful sky-blue colour. Or it may become red, or occasionally yellow. All of these troubles the dairyman owes to the presence in his milk of unusual species of bacteria which grow there abundantly.

18. The word *deleterious* in the first paragraph is most closely related to the word _____
    a. amicable.
    b. smoldering.
    c. luminous.
    d. ruinous.

19. Which of the following best explains how the passage is organized?
    a. The author begins by presenting the effects of a phenomenon, then explains the process of this phenomenon, and then ends by giving the history of the study of this phenomenon.
    b. The author begins by explaining a process or phenomenon, then gives the history of the study of this phenomenon, then ends by presenting the effects of this phenomenon.
    c. The author begins by giving the history of the study of a certain phenomenon, then explains the process of this phenomenon, then ends by presenting the effects of this phenomenon.
    d. The author begins by giving a broad definition of a subject, then presents more specific cases of the subject, then ends by contrasting two different viewpoints on the subject.

20. The primary purpose of the passage is _____

    a. to inform the reader of the phenomenon, investigation, and consequences of milk souring.

    b. to persuade the reader that milk souring is due to *Bacillus acidi lactici*, found commonly in the dairies of Europe.

    c. to describe the accounts and findings of researchers studying the phenomenon of milk souring.

    d. to discount the former researchers' opinions on milk souring and bring light to new investigations.

21. The author says _____ about milk souring.

    a. Milk souring is caused mostly by a species of bacteria called *Bacillus acidi lactici*, although former research asserted that it was caused by a variety of bacteria.

    b. The ordinary cause of milk souring is unknown to current researchers, although former researchers thought it was due to a species of bacteria called *Bacillus acidi lactici*.

    c. Milk souring is caused mostly by a species of bacteria identical to that of *Bacillus acidi lactici*, although there are a variety of other bacteria that cause milk souring as well.

    d. The ordinary cause of milk souring will sometimes curdle without becoming acidic, though sometimes it will turn colors other than white, or have strange smells or tastes.

22. The author of the passage would most likely agree with the fact that _____

    a. milk researchers in the past have been incompetent and have sent us on a wild goose chase when determining what causes milk souring.

    b. dairymen are considered more expert in the field of milk souring than milk researchers.

    c. the study of milk souring has improved throughout the years, as we now understand more of what causes milk souring and what happens afterward.

    d. any type of bacteria will turn milk sour, so it's best to keep milk in an airtight container while it is being used.

23. Given the author's account of the consequences of milk souring, _____ is most closely analogous to the author's description of what happens after milk becomes slimy?

    a. the chemical change that occurs when a firework explodes

    b. a rainstorm that overwaters a succulent plant

    c. mercury inside of a thermometer that leaks out

    d. a child who swallows flea medication

24. _____ would most likely come after the third paragraph.

    a. A paragraph depicting the general effects of bacteria on milk

    b. A paragraph explaining a broad history of what researchers have found in regard to milk souring

    c. A paragraph outlining the properties of milk souring and the way in which it occurs

    d. A paragraph showing the ways bacteria infiltrate milk and ways to avoid this infiltration

The next three questions are based on the following passage from The Life, Crime, and Capture of John Wilkes Booth *by George Alfred Townsend.*

> Having completed these preparations, Mr. Booth entered the theater by the stage door; summoned one of the scene shifters, Mr. John Spangler, emerged through the same door with that individual, leaving the door open, and left the mare in his hands to be held until he (Booth) should return. Booth who was even more fashionably and richly dressed than usual, walked thence around to the front of the theater, and went in. Ascending to the dress circle, he stood for a little time gazing around upon the audience and occasionally upon the stage in his usual graceful manner. He was subsequently observed by Mr. Ford, the proprietor of the theater, to be slowly elbowing his way through the crowd that packed the rear of the dress circle toward the right side, at the extremity of which was the box where Mr. and Mrs. Lincoln and their companions were seated. Mr. Ford casually noticed this as a slightly extraordinary symptom of interest on the part of an actor so familiar with the routine of the theater and the play.

25. The above passage is organized by _____
    a. chronology.
    b. cause and effect.
    c. problem to solution.
    d. main idea with supporting details.

26. Based on your knowledge of history, _____
    a. an asteroid is about to hit the earth.
    b. the best opera of all times is about to premiere.
    c. a playhouse is about to be burned to the ground.
    d. a president is about to be assassinated.

27. What does the author mean by the last two sentences?
    a. Mr. Ford was suspicious of Booth and assumed he was making his way to Mr. Lincoln's box.
    b. Mr. Ford assumed Booth's movement throughout the theater was due to being familiar with the theater.
    c. Mr. Ford thought that Booth was making his way to the theater lounge to find his companions.
    d. Mr. Ford thought that Booth was elbowing his way to the dressing room to get ready for the play.

28. A document explaining how to use an LCD flat screen TV is known as a _____ document?
    a. technical
    b. persuasive
    c. narrative
    d. cause and effect

The next four questions are based on the following passage. It is from Oregon, Washington, and Alaska. Sights and Scenes for the Tourist, *written by E.L. Lomax in 1890:*

> Portland is a very beautiful city of 60,000 inhabitants, and situated on the Willamette river twelve miles from its junction with the Columbia. It is perhaps true of many of the growing cities of the West, that they do not offer the same social advantages as the older cities of the East. But this is principally the case as to what may be called boom cities, where the larger part of the population is of that floating class which follows in the line of temporary growth for the purposes of speculation, and in no sense applies to those centers of trade whose prosperity is based on the solid foundation of legitimate business. As the metropolis of a vast section of country, having

broad agricultural valleys filled with improved farms, surrounded by mountains rich in mineral wealth, and boundless forests of as fine timber as the world produces, the cause of Portland's growth and prosperity is the trade which it has as the center of collection and distribution of this great wealth of natural resources, and it has attracted, not the boomer and speculator, who find their profits in the wild excitement of the boom, but the merchant, manufacturer, and investor, who seek the surer if slower channels of legitimate business and investment. These have come from the East, most of them within the last few years. They came as seeking a better and wider field to engage in the same occupations they had followed in their Eastern homes, and bringing with them all the love of polite life which they had acquired there, have established here a new society, equaling in all respects that which they left behind. Here are as fine churches, as complete a system of schools, as fine residences, as great a love of music and art, as can be found at any city of the East of equal size.

29. A "boom city" is _____
   a. a city that is built on solid business foundation of mineral wealth and farming.
   b. an area of land on the west coast that quickly becomes populated by residents from the east coast.
   c. a city that, due to the hot weather and dry climate, catches fire frequently, resulting in a devastating population drop.
   d. a city whose population is made up of people who seek quick fortunes rather than building a solid business foundation.

30. According to the author, Portland is _____
   a. a boom city.
   b. a city on the East Coast.
   c. an industrial city.
   d. a city of legitimate business.

31. This passage is _____
   a. a business proposition.
   b. a travel guide.
   c. a journal entry.
   d. a scholarly article.

The following two questions are based on the excerpt from The Golden Bough *by Sir James George Frazer.*

   The other of the minor deities at Nemi was Virbius. Legend had it that Virbius was the young Greek hero Hippolytus, chaste and fair, who learned the art of venery from the centaur Chiron, and spent all his days in the greenwood chasing wild beasts with the virgin huntress Artemis (the Greek counterpart of Diana) for his only comrade.

32. Based on a prior knowledge of literature, the reader can infer this passage is taken from _____
   a. a eulogy.
   b. a myth.
   c. a historical document.
   d. a technical document.

33. The meaning of the word "comrade" as the last word in the passage is ____
    a. friend.
    b. enemy.
    c. brother.
    d. pet.

The next two questions are based on the following passage from Walden *by Henry David Thoreau.*

> When I wrote the following passages, or rather the bulk of them, I lived alone, in the woods, a mile from any neighbor, in a house which I had built myself on the shore of Walden Pond, in Concord, Massachusetts, and earned my living by the labor of my hands only. I lived there two years and two months. At present I am a sojourner in civilized life again.

34. The text is most likely to be found in a(n) _____
    a. introduction.
    b. appendix.
    c. dedication.
    d. glossary.

35. The word *sojourner* most likely means ____
    a. illegal alien.
    b. temporary resident.
    c. lifetime partner.
    d. farm crop.

The next two questions are based on the following passage, which is a preface for Poems by Alexander Pushkin *by Ivan Panin.*

> I do not believe there are as many as five examples of deviation from the literalness of the text. Once only, I believe, have I transposed two lines for convenience of translation; the other deviations are (*if* they are such) a substitution of an *and* for a comma in order to make now and then the reading of a line musical. With these exceptions, I have sacrificed *everything* to faithfulness of rendering. My object was to make Pushkin himself, without a prompter, speak to English readers. To make him thus speak in a foreign tongue was indeed to place him at a disadvantage; and music and rhythm and harmony are indeed fine things, but truth is finer still. I wished to present not what Pushkin would have said, or [Pg 10] should have said, if he had written in English, but what he does say in Russian. That, stripped from all ornament of his wonderful melody and grace of form, as he is in a translation, he still, even in the hard English tongue, soothes and stirs, is in itself a sign that through the individual soul of Pushkin sings that universal soul whose strains appeal forever to man, in whatever clime, under whatever sky.

36. From clues in this passage, the author is doing ____
    a. translation work.
    b. criticism.
    c. historical validity.
    d. a biography.

37. According to the author, _____ is the most important aim of translation work.
    a. to retain the beauty of the work
    b. to retain the truth of the work
    c. to retain the melody of the work
    d. to retain the form of the work

The following four questions are based on the excerpt from Variation of Animals and Plants *by Charles Darwin.*

> Peach (Amygdalus persica).—In the last chapter I gave two cases of a peach-almond and a double-flowered almond which suddenly produced fruit closely resembling true peaches. I have also given many cases of peach-trees producing buds, which, when developed into branches, have yielded nectarines. We have seen that no less than six named and several unnamed varieties of the peach have thus produced several varieties of nectarine. I have shown that it is highly improbable that all these peach-trees, some of which are old varieties, and have been propagated by the million, are hybrids from the peach and nectarine, and that it is opposed to all analogy to attribute the occasional production of nectarines on peach-trees to the direct action of pollen from some neighbouring nectarine-tree. Several of the cases are highly remarkable, because, firstly, the fruit thus produced has sometimes been in part a nectarine and in part a peach; secondly, because nectarines thus suddenly produced have reproduced themselves by seed; and thirdly, because nectarines are produced from peach-trees from seed as well as from buds. The seed of the nectarine, on the other hand, occasionally produces peaches; and we have seen in one instance that a nectarine-tree yielded peaches by bud-variation. As the peach is certainly the oldest or primary variety, the production of peaches from nectarines, either by seeds or buds, may perhaps be considered as a case of reversion. Certain trees have also been described as indifferently bearing peaches or nectarines, and this may be considered as bud-variation carried to an extreme degree.

38. _____ is NOT a detail from the passage.
    a. At least six named varieties of the peach have produced several varieties of nectarine.
    b. It is not probable that all of the peach-trees mentioned are hybrids from the peach and nectarine.
    c. An unremarkable case is the fact that nectarines are produced from peach-trees from seed as well as from buds.
    d. The production of peaches from nectarines might be considered a case of reversion.

39. The author's tone in this passage is _____
    a. enthusiastic.
    b. objective.
    c. critical.
    d. desperate.

40. Which of the following is an accurate paraphrasing of the following phrase?

Certain trees have also been described as indifferently bearing peaches or nectarines, and this may be considered as bud-variation carried to an extreme degree.

    a. Some trees are described as bearing peaches and some trees have been described as bearing nectarines, but individually the buds are extreme examples of variation.
    b. One way in which bud-variation is said to be carried to an extreme degree is when specific trees have been shown to casually produce peaches or nectarines.
    c. Certain trees are indifferent to bud-variation, as recently shown in the trees that produce both peaches and nectarines in the same season.
    d. Nectarines and peaches are known to have cross-variation in their buds, which indifferently bears other sorts of fruit to an extreme degree.

The following four questions are based on the excerpt from A Christmas Carol *by Charles Dickens.*

> Meanwhile the fog and darkness thickened so, that people ran about with flaring links, proffering their services to go before horses in carriages, and conduct them on their way. The ancient tower of a church, whose gruff old bell was always peeping slily down at Scrooge out of a Gothic window in the wall, became invisible, and struck the hours and quarters in the clouds, with tremulous vibrations afterwards as if its teeth were chattering in its frozen head up there. The cold became intense. In the main street, at the corner of the court, some labourers were repairing the gas-pipes, and had lighted a great fire in a brazier, round which a party of ragged men and boys were gathered: warming their hands and winking their eyes before the blaze in rapture. The water-plug being left in solitude, its overflowings sullenly congealed, and turned to misanthropic ice. The brightness of the shops where holly sprigs and berries crackled in the lamp heat of the windows, made pale faces ruddy as they passed. Poulterers' and grocers' trades became a splendid joke; a glorious pageant, with which it was next to impossible to believe that such dull principles as bargain and sale had anything to do. The Lord Mayor, in the stronghold of the mighty Mansion House, gave orders to his fifty cooks and butlers to keep Christmas as a Lord Mayor's household should; and even the little tailor, whom he had fined five shillings on the previous Monday for being drunk and bloodthirsty in the streets, stirred up to-morrow's pudding in his garret, while his lean wife and the baby sallied out to buy the beef.

41. In the context in which it appears, *congealed* most nearly means ____
    a. burst.
    b. loosened.
    c. shrank.
    d. thickened.

42. The statement that _____ can NOT be inferred from the passage.
    a. the season of this narrative is in the wintertime
    b. the majority of the narrative is located in a bustling city street
    c. this passage takes place during the nighttime
    d. the Lord Mayor is a wealthy person within the narrative

43. According to the passage, the poulterers and grocers _____
    a. were so poor in the quality of their products that customers saw them as a joke.
    b. put on a pageant in the streets every year for Christmas to entice their customers.
    c. did not believe in Christmas so they refused to participate in the town parade.
    d. set their shops up to be entertaining public spectacles rather than a dull trade exchange.

44. The meaning of the word "proffering" in this passage is _____
    a. giving away.
    b. offering.
    c. bolstering.
    d. teaching.

45. Journalist: Our newspaper should only consider the truth in its reporting. When a party is clearly in the wrong, like if he or she is spreading a pernicious, false narrative, their position should never be presented alongside the truth without comment. The purpose of journalism is to deliver facts and context. Both sides of an issue should be called for comment, but their responses should be framed appropriately, especially when there's a potential conflict of interest or source of bias at play. Our editorial board needs to seriously consider how our newspaper isn't currently meeting these basic standards, exposing us to charges of bias from all sides.

The primary purpose of the journalist's argument is to _____
    a. persuade the newspaper to adopt a more rigorous approach to journalism.
    b. defend the newspaper against charges of bias in its reporting.
    c. argue for the newspaper to hire more journalists with the appropriate skills.
    d. define the professional responsibilities of a journalist.

For the next question, select the choice you think best fits the underlined part of the sentence. If the original is the best answer choice, then choose Choice A.

46. Play baseball, swimming, and dancing are three of Hannah's favorite ways to be active.
    a. Play baseball, swimming, and dancing
    b. Playing baseball; swimming; dancing;
    c. Playing baseball, to swim and to dance,
    d. Playing baseball, swimming, and dancing

The next question is based on the following passage from The Federalist No. 78 *by Alexander Hamilton.*

> According to the plan of the convention, all judges who may be appointed by the United States are to hold their offices *during good behavior*, which is conformable to the most approved of the State constitutions and among the rest, to that of this State. Its propriety having been drawn into question by the adversaries of that plan, is no light symptom of the rage for objection, which disorders their imaginations and judgments. The standard of good behavior for the continuance in office of the judicial magistracy, is certainly one of the most valuable of the modern improvements in the practice of government. In a monarchy it is an excellent barrier to the despotism of the prince; in a republic it is a no less excellent barrier to the encroachments and oppressions of the representative body. And it is the best expedient which can be devised in any government, to secure a steady, upright, and impartial administration of the laws.

47. Hamilton's point in this passage is _____
    a. to show the audience that despotism within a monarchy is no longer the standard practice in the states.
    b. to convince the audience that judges holding their positions based on good behavior is a practical way to avoid corruption.
    c. to persuade the audience that having good behavior should be the primary characteristic of a person in a government body and their voting habits should reflect this.
    d. to convey the position that judges who serve for a lifetime will not be perfect and therefore we must forgive them for their bad behavior when it arises.

48. Maritza wanted to go to the park to swim with her friends, but when she got home, she realized that nobody was there to take her.

Follow the numbered instructions to transform the sentence above into a new sentence.
    1. Replace the phrase "wanted to go" with "went."
    2. Replace the word "but" with the word "and."
    3. Replace the word "home" with "there."
    4. Replace the phrase "nobody was there" with "they had thrown."
    5. Take out the phrase "to take."
    6. Add the words "a surprise party!" at the end of the sentence.

    a. Maritza went to the park to swim with her friends, and when she got home, she realized they had thrown her a surprise party!
    b. Maritza wanted to go to the park to swim with her friends, but when she got home, she realized they had thrown her a surprise party!
    c. Maritza went to the park to swim with her friends, and when she got there, she realized that they had thrown her a surprise party!
    d. Maritza went to the park to swim with her friends, and when she got there, she realized that nobody had thrown her a surprise party!

49. A reader comes across a word they do not know in the book they are reading, and they need to find out what the word means in order to understand the context of the sentence. The reader should look in the ____
    a. table of contents.
    b. introduction.
    c. index.
    d. glossary.

The next three questions are based on the following passage from The Biography of Queen Victoria *by E. Gordon Browne, M.A.:*

The old castle soon proved to be too small for the family, and in September 1853 the foundation-stone of a new house was laid. After the ceremony the workmen were entertained at dinner, which was followed by Highland games and dancing in the ballroom.

Two years later they entered the new castle, which the Queen described as "charming; the rooms delightful; the furniture, papers, everything perfection."

The Prince was untiring in planning improvements, and in 1856 the Queen wrote: "Every year my heart becomes more fixed in this dear Paradise, and so much more so now, that *all* has become my dearest Albert's *own* creation, own work, own building, own laying out as at Osborne; and his great taste, and the impress of his dear hand, have been stamped everywhere. He was very busy today, settling and arranging many things for next year."

50. This excerpt is considered a _____
    a. primary source.
    b. secondary source.
    c. tertiary source.
    d. None of these

51. The castle took _____ to be built.
    a. one year
    b. two years
    c. three years
    d. four years

52. The word *impress* means _____
    a. to affect strongly in feeling.
    b. to urge something to be done.
    c. to impose a certain quality upon.
    d. to press a thing onto something else.

The following passage is taken from Chapter 6 of Sense and Sensibility, *by Jane Austen:*

As a house, Barton Cottage, though small, was comfortable and compact; but as a cottage it was defective, for the building was regular, the roof was tiled, the window shutters were not painted green, nor were the walls covered with honeysuckles. A narrow passage led directly through the house into the garden behind. On each side of the entrance was a sitting room, about sixteen feet square; and beyond them were the offices and the stairs. Four bed-rooms and two garrets formed the rest of the house. It had not been built many years and was in good repair. In comparison of Norland, it was poor and small indeed!—but the tears which recollection called forth as they entered the house were soon dried away. They were cheered by the joy of the servants on their arrival, and each for the sake of the others resolved to appear happy. It was very early in September; the season was fine, and from first seeing the place under the advantage of good weather, they received an impression in its favour which was of material service in recommending it to their lasting approbation.

53. This passage is describing a _____
    a. museum.
    b. cottage.
    c. skyscraper.
    d. island.

54. The narrow passage led through the house into the _____
    a. office.
    b. bedroom.
    c. kitchen.
    d. garden.

55. _____ was on each side of the entrance.
    a. A sitting room
    b. A cat
    c. A statue
    d. A dining room

56. There were _____ bedrooms inside the house.
    a. two
    b. three
    c. four
    d. five

57. The passage is set in the month of _____
    a. April.
    b. October.
    c. September.
    d. December.

The following five questions are based on the following passage:

> Smart assistants like "Google Home" and "Amazon Echo" are great additional features to have in the home. You can buy these assistants online for a couple hundred dollars from major retailers throughout the country. These smart assistants connect wirelessly to other electronic devices in your home and act on command to make your life more efficient. For example, you can say "Alexa, turn the music up forty percent," and the music goes up. Or you can say, "OK Google, dim the lights," and your lights will dim without you having to step a foot off the couch! Devices hooked up to your TV also follow commands, such as searching for certain TV shows or playing your favorite selections.
>
> Although many people believe otherwise, there is really no need to worry about safety with smart assistants; we are told by people like Janie Adams and Ray Gold, both trustworthy CEO's of big tech companies, that a small percentage of our information goes to advertisers, but there is a limit to what and how much is sent to them. What's the big deal, anyway, if the shoes you have been dreaming about pop up on your google search every so often? There is no harm here because we don't have any proof that says the government has our information and is using it to monitor us.

58. The topic of the passage is about _____
    a. privacy.
    b. smart assistants.
    c. government.
    d. furniture.

59. According to the passage, smart devices _____
    a. make your life more efficient.
    b. are not worth the money.
    c. are incredibly dangerous.
    d. are very complicated to set up.

60. Janie Adams and Ray Gold are _____
    a. historical figures in the Civil Rights Movement.
    b. authors of technical manuals.
    c. inventors of smart devices.
    d. CEO's of big tech companies.

61. According to the passage, _____ percentage of our information goes to advertisers.
    a. zero
    b. a small
    c. a medium
    d. a large

62. This passage is considered _____ writing.
    a. informative
    b. narrative
    c. descriptive
    d. persuasive

Answer Explanations

1. A: To show the audience one of the effects of criminal rehabilitation by comparison. Choice *B* is incorrect because although it is obvious the author favors rehabilitation, the author never asks for donations from the audience. Choices *C* and *D* are also incorrect. We can infer from the passage that American prisons are probably harsher than Norway prisons. However, the best answer that captures the author's purpose is Choice *A*, because we see an effect by the author (recidivism rate of each country) comparing Norwegian and American prisons.

2. D: The likelihood of a convicted criminal to reoffend. The passage explains how a Norwegian prison, due to rehabilitation, has a smaller rate of recidivism. Thus, we can infer that recidivism is probably not a positive attribute. Choices *A* and *B* are both positive attributes, the lack of violence and the opportunity of inmates to receive therapy, so Norway would probably not have a lower rate of these two things. Choice *C* is possible, but it does not make sense in context, because the author does not talk about tactics in which to keep prisoners inside the compound, but ways in which to rehabilitate criminals so that they can live as citizens when they get out of prison.

3. C: The passage is reflective of a narrative. A narrative is used to tell a story, as we see the narrator trying to do in this passage by using memory and dialogue. Persuasive writing uses rhetorical devices to try and convince the audience of something, and there is no persuasion or argument within this passage. Expository writing is a type of writing used to inform the reader. Technical writing is usually used within business communications and uses technical language to explain procedures or concepts to someone within the same technical field.

4. C: A sense of foreboding. The narrator, after feeling excitement for the morning, feels "that something awful was about to happen," which is considered foreboding. The narrator mentions larks and weather in the passage, but there is no proof of anger or confusion at either of them.

5. C: To introduce Peter Walsh back into the narrator's memory. Choice *A* is incorrect because, although the novel *Mrs. Dalloway* is about events leading up to a party, the passage does not mention anything about a party. Choice *B* is incorrect; the narrator calls Peter *dull* at one point, but the rest of her memories of him are more positive. Choice *D* is incorrect; although morning is described within the first few sentences of the passage, the passage quickly switches to a description of Peter Walsh and the narrator's memories of him.

6. C: Personification. Figurative language uses words or phrases that are different from their literal interpretation. Personification is included in figurative speech, which means giving inanimate objects human characteristics. Nouns, agreement, and syntax all have to do with grammar and usage, and are not considered figurative language.

7. D: Felicia had to be prudent, or careful, if she was going to cross the bridge over the choppy water. Choice *A*, patient, is close to the word careful. However, careful makes more sense here. Choices *B* and *C* don't make sense within the context—Felicia wasn't hoping to be *afraid* or *dangerous* while crossing over the bridge, but was hoping to be careful to avoid falling.

8. B: Whatever happened in his life before he had a certain internal change is irrelevant. Choices *A, C,* and *D* use some of the same language as the original passage, like "revolution," "speak," and "details," but they do not capture the meaning of the statement. The statement is saying the details of his previous life are not going to be talked about—that he had some kind of epiphany, and moving forward in his life is what the narrator cares about.

9. B: First-person omniscient. This is the best guess with the information we have. In the world of the passage, the narrator is first-person, because we see them use the "I," but they also know the actions and thoughts of the protagonist, a character named "Webster." First-person limited tells their own story, making Choice *A* incorrect. Choice *C* is incorrect; second person uses "you" to tell the story. Third person uses "them," "they," etc., and would not fall into use of the "I" in the narrative, making Choice *D* incorrect.

10. A: Webster is a washing machine manufacturer. This question depends on reading comprehension. We see in the second sentence that Webster "was a fairly prosperous manufacturer of washing machines," making Choice A the correct answer.

11. B: Veggie pasta with marinara sauce. Choices *A* and *D* are incorrect because they both contain cheese, and the doctor gave Sheila a list *without* dairy products. Choice *C* is incorrect because the doctor is also having Sheila stay away from eggs, and the omelet has eggs in it. Choice *B* is the best answer because it contains no meat, dairy, or eggs.

12. C: Choice *C* correctly identifies a necessary assumption in the argument. The argument is that hunting licenses for endangered species should be sold to support conservation efforts. If the revenue from those licenses isn't funding conservation efforts, then the ecologist's entire argument falls apart.

Choice *A* is incorrect. Hunting licenses for non-endangered species could be profitable, and the ecologist's argument wouldn't be impacted one way or another. The ecologist could still argue that hunting licenses should be expanded to further increase funding for species that are still endangered.

Choice *B* is incorrect. The ecologist would definitely agree that all one hundred animal species should be saved, but it isn't a necessary assumption. The argument would function the same if the ultimate goal were to save ten endangered animal species.

Choice *D* is incorrect. Like Choice *B*, the ecologist would agree that conservation efforts should've begun earlier, but it isn't a necessary assumption.

13. B: Choice *B* is the best answer here; the sentence states "In unknown regions take a responsible guide with you, unless the trail is short, easily followed, and a frequented one." Choice *A* is incorrect; the passage does not state that you should try and explore unknown regions. Choice *C* is incorrect; the passage talks about trails that contain pitfalls, traps, and boggy places, but it does not say that *all* unknown regions contain these things. Choice *D* is incorrect; the passage mentions "rail" and "boat" as means of transport at the beginning, but it does not suggest it is better to travel unknown regions by rail.

14. D: Choice *D* is correct; it may be real advice an experienced hiker would give to an inexperienced hiker. However, the question asks about details in the passage, and this is not in the passage. Choice *A* is incorrect; we do see the author encouraging the reader to learn about the trail beforehand . . . "wet or dry; where it leads; and its length." Choice *B* is also incorrect, because we do see the author telling us the time will lengthen with boggy or rugged places opposed to smooth places. Choice *C* is incorrect; at the end of the passage, the author tells us "do not go alone through lonely places . . . unless you are quite familiar with the country and the ways of the wild."

15. D: This is an informative passage. Informative passages explain to the readers how to do something; in this case, the author is attempting to explain the fundamentals of camping and hiking. Descriptive is a type of passage describing a character, event, or place in great detail and imagery, so this is incorrect. A persuasive passage is an argument that tries to get readers to agree with something. A narrative is a passage that tells a story, so this is also incorrect.

16. B: Choice *B* correctly identifies a statement that the ethicist would agree with. The ethicist is troubled by the lack of incentives for private companies to regulate artificial intelligence. According to the ethicist's conclusion, artificial intelligence is adopting humanity's worst impulses and violent worldviews, and thereby threatening humanity.

Choice *A* is incorrect. The ethicist might agree that artificial intelligence is negatively impacting society, but the Internet is barely mentioned in this argument. It's unclear whether the ethicist believes that the Internet's influence on artificial intelligence makes the Internet have a negative impact on society.

Choice *C* is incorrect. Although the ethicist would likely agree with this general sentiment, Choice *C* is too broad. The argument is limited to artificial intelligence, not technological advancement generally, and it's unclear whether the ethicist would *always* take this position in those additional scenarios.

Choice *D* is incorrect. The ethicist doesn't argue for limiting artificial intelligence to simple algorithms. The problem isn't the advancement; it's that the artificial intelligence's growth isn't aligned with humanity's interest, according to the ethicist.

17. B: Choice *B* correctly identifies the best investment opportunity. The cardinal rule has three requirements—privately held small business, sells tangible goods, and no debt. Elizabeth's store is a privately held small business (standalone and owned by her), it sells tangible goods (board games), and it has no debt. The lack of profitability is irrelevant, acting as a red herring. The cardinal rule doesn't mention it, presumably since the businessman thinks he can increase profitability as long as the business meets those three requirements.

Choice *A* is incorrect. Jose's grocery store owes the bank three months' worth of mortgage payments, so it has debt, violating the cardinal rule.

Choice *C* is incorrect. The accounting firm violates the cardinal rule, because it does not sell a tangible good.

Choice *D* is incorrect. A multinational corporation is not a small business, so it violates the cardinal rule.

18. D: The word *deleterious* can be best interpreted as referring to the word *ruinous*. The first paragraph attempts to explain the process of milk souring, so the "acid" would probably prove "ruinous" to the growth of bacteria and cause souring. Choice *A, amicable*, means friendly, so this does not make sense in context. Choice *B, smoldering*, means to boil or simmer, so this is also incorrect. Choice *C, luminous*, has positive connotations and doesn't make sense in the context of the passage. Luminous means shining or brilliant.

19. B: The author begins by explaining a process or phenomenon, then gives the history of the study of this phenomenon, then ends by presenting the effects of this phenomenon. The author explains the process of souring in the first paragraph by informing the reader that "it is due to the action of certain of the milk bacteria upon the milk sugar which converts it into lactic acid, and this acid gives the sour taste and curdles the milk." In the second paragraph, we see how the phenomenon of milk souring was viewed when it was "first studied," and then we proceed to gain insight into "recent investigations" toward the end of the paragraph. Finally, the passage ends by presenting the effects of the phenomenon of milk souring. We see the milk curdling, becoming bitter, tasting soapy, turning blue, or becoming thread-like. All of the other answer choices are incorrect.

20: A: To inform the reader of the phenomenon, investigation, and consequences of milk souring. Choice *B* is incorrect because the passage states that *Bacillus acidi lactici* is not the only cause of milk souring. Choice *C* is incorrect because, although the author mentions the findings of researchers, the main purpose of the text does not seek to describe their accounts and findings, as we are not even told the names of any of the researchers. Choice *D* is tricky. We do see the author present us with new findings in contrast to the first cases studied by researchers. However, this information is only in the second paragraph, so it is not the primary purpose of the *entire passage*.

21. C: Milk souring is caused mostly by a species of bacteria identical to that of *Bacillus acidi lactici* although there are a variety of other bacteria that cause milk souring as well. Choice *A* is incorrect because it contradicts the assertion that the souring is still caused by a variety of bacteria. Choice *B* is incorrect because the ordinary cause of milk souring *is known* to current researchers. Choice *D* is incorrect because this names mostly the effects of milk souring, not the cause.

22. C: The study of milk souring has improved throughout the years, as we now understand more of what causes milk souring and what happens afterward. None of the choices here are explicitly stated, so we have to rely on our ability to make inferences. Choice *A* is incorrect because there is no indication from the author that milk researchers in the past have been incompetent—only that recent research has done a better job of studying the phenomenon of milk souring. Choice *B* is incorrect because the author refers to dairymen in relation to the effects of milk souring and their "troubles" surrounding milk souring, and does not compare them to milk researchers. Choice *D* is incorrect because we are told in the second paragraph that only certain types of bacteria are able to sour milk. Choice *C* is the best answer choice here because although the author does not directly state that the study of milk souring has improved, we can see this might be true due to the comparison of old studies to newer studies, and the fact that the newer studies are being used as a reference in the passage.

23. A: The chemical change that occurs when a firework explodes. The author tells us that after milk becomes slimy, "it persists in spite of all attempts made to remedy it," which means the milk has gone through a chemical change. It has changed its state from milk to sour milk by changing its odor, color, and material. After a firework explodes, there is nothing one can do to change the substance of a firework back to its original form—the original substance is turned into sound and light. Choice *B* is incorrect because, although the rain overwatered the plant, it's possible that the plant is able to recover from this. Choice *C* is incorrect because although Mercury leaking out may be dangerous, the actual substance itself stays the same and does not alter into something else. Choice *D* is incorrect; this situation is not analogous to the alteration of a substance.

24. D: A paragraph showing the ways bacteria infiltrate milk and ways to avoid this infiltration. Choices *A, B,* and *C* are incorrect because these are already represented in the third, second, and first paragraphs. Choice *D* is the best answer because it follows a sort of problem/solution structure in writing.

25. A: The passage is organized by chronology. The passage presents us with a sequence of events that happens in chronological order. Choice *B* is incorrect. Cause and effect organization would usually explain why something happened or list the effects of something. Choice *C* is incorrect because problem and solution organization would detail a problem and then present a solution to the audience, and there is no solution presented here. Finally, Choice *D* is incorrect. We are entered directly into the narrative without any main idea or any kind of argument being delivered.

26. D: A president is about to be assassinated. The context clues in the passage give hints to what is about to happen. The passage mentions John Wilkes Booth as "Mr. Booth," the man who shot Abraham Lincoln. The passage also mentions a "Mr. Ford," and we know that Lincoln was shot in Ford's theater. Finally, the passage mentions Mr. and Mrs. Lincoln. By adding all these clues up and our prior knowledge of history, the assassination of President Lincoln by Booth in Ford's theater is probably the next thing that is going to happen.

27. B: Mr. Ford assumed Booth's movement throughout the theater was due to being familiar with the theater. Choice *A* is incorrect; although Booth does eventually make his way to Lincoln's box, Mr. Ford does not make this distinction in this part of the passage. Choice *C* is incorrect; although the passage mentions "companions," it mentions Lincoln's companions rather than Booth's companions. Finally, Choice *D* is incorrect; the passage mentions "dress circle," which means the first level of the theater, but this is different from a "dressing room."

28. A: Technical document. Technical documents are documents that describe the functionality of a technical product, so Choice *A* is the best answer. Persuasive texts, Choice *B*, try to persuade an audience to follow the author's line of thinking or to act on something. Choice *C*, narrative texts, seek to tell a story. Choice *D*, cause and effect, try to show why something happened, or the causes or effects of a particular thing.

29. D: A city whose population is made up of people who seek quick fortunes rather than building a solid business foundation. Choice *A* is a characteristic of Portland, but not that of a boom city. Choice *B* is close—a boom city is one that becomes quickly populated, but it is not necessarily always populated by residents from the east coast. Choice *C* is incorrect because a boom city is not one that catches fire frequently, but one made up of people who are looking to make quick fortunes from the resources provided on the land.

30. D: A city of legitimate business. We can see the proof in this sentence: "the cause of Portland's growth and prosperity is the trade which it has as the center of collection and distribution of this great wealth of natural resources, and it has attracted, not the boomer and speculator . . . but the merchant, manufacturer, and investor, who seek the surer if slower channels of legitimate business and investment." Choices *A, B,* and *C* are not mentioned in the passage and are incorrect.

31. B: A travel guide. Our first hint is in the title: *Oregon, Washington, and Alaska. Sights and Scenes for the Tourist.* Although the passage talks about business, there is no proposition included, which makes Choice *A* incorrect. Choice *C* is incorrect because the style of the writing is more informative and formal rather than personal and informal. Choice *D* is incorrect; this could possibly be a scholarly article, but the best choice is that it is a travel guide, due to the title and the details of what the city has to offer at the very end.

32. B: A myth. Look for the key words that give away the type of passage this is, such as "deities," "Greek hero," "centaur," and the names of demigods like Artemis. A eulogy is typically a speech given at a funeral, making Choice *A* incorrect. Choices *C* and *D* are incorrect, as "virgin huntresses" and "centaurs" are typically not found in historical or professional documents.

33. A: Friend. Based on the context of the passage, we can see that Hippolytus was a friend to Artemis because he "spent all his days in the greenwood chasing wild beasts" with her.

34. A: Introduction. The passage tells us how the following passages of the book are written, which acts as an introduction to the work. An appendix comes at the end of a book, giving extra details, making

Choice *B* incorrect. Choice *C* is incorrect; a dedication is at the beginning of a book, but usually acknowledges those the author is grateful towards. Choice *D* is incorrect because a glossary is a list of terms with definitions at the end of a chapter or book.

35. B: Temporary resident. Although we don't have much context to go off of, we know that one is probably not a "lifetime partner" or "farm crop" of civilized life. These two do not make sense, so Choices *C* and *D* are incorrect. Choice *A* is also a bit strange. To be an "illegal alien" of civilized life is not a used phrase, making Choice *A* incorrect.

36. A: The author is doing translation work. We see this very clearly in the way the author talks about staying truthful to the original language of the text. The text also mentions "translation" towards the end. Criticism is taking an original work and analyzing it, making Choice *B* incorrect. The work is not being tested for historical validity, but being translated into the English language, making Choice *C* incorrect. The author is not writing a biography, as there is nothing in here about Pushkin himself, only his work, making Choice *D* incorrect.

37. B: To retain the truth of the work. The author says that "music and rhythm and harmony are indeed fine things, but truth is finer still," which means that the author stuck to a literal translation instead of changing up any words that might make the English language translation sound better.

38. C: This question requires close attention to the passage. Choice *A* can be found where the passage says "no less than six named and several unnamed varieties of the peach have thus produced several varieties of nectarine," so this choice is incorrect. Choice *B* can be found where the passage says "it is highly improbable that all these peach-trees . . . are hybrids from the peach and nectarine." Choice *D* is incorrect because we see in the passage that "the production of peaches from nectarines, either by seeds or buds, may perhaps be considered as a case of reversion." Choice *C* is the correct answer because the word "unremarkable" should be changed to "remarkable" in order for it to be consistent with the details of the passage.

39. B: The author's tone in this passage can be considered objective. An objective tone means that the author is open-minded and detached about the subject. Most scientific articles are objective. Choices *A*, *C*, and *D* are incorrect. The author is not very enthusiastic on the paper; the author is not critical, but rather interested in the topic. The author is not desperate in any way here.

40. B: Choice *B* is the correct answer because the meaning holds true even if the words have been switched out or rearranged some. Choice *A* is incorrect because it has trees either bearing peaches or nectarines, and the trees in the original phrase bear both. Choice *C* is incorrect because the statement does not say these trees are "indifferent to bud-variation," but that they have "indifferently [bore] peaches or nectarines." Choice *D* is incorrect; the statement may use some of the same words, but the meaning is skewed in this sentence.

41. D: *Congealed* in this context most nearly means *thickened*, because we see liquid turning into ice. Choice *A*, Choice *B*, *loosened*, is the opposite of the correct answer. Choices *A* and *C*, burst and shrank, are also incorrect.

42. C: Choice *C* is correct. We cannot infer that the passage takes place during the nighttime. While we do have a statement that says that the darkness thickened, this is the only evidence we have. The darkness could be thickening because it is foggy outside. We don't have enough proof to infer this otherwise. We *can* infer that the season of this narrative is in the wintertime. Some of the evidence here is that "the cold became intense," and people were decorating their shops with "holly sprigs," a Christmas tradition. It also mentions that it's Christmas time at the end of the passage. Choice *B* is incorrect; we *can* infer that the narrative is located in a bustling city street by the actions in the story. People are running around trying to sell things, the atmosphere is busy, there is a church tolling the hours, etc. The scene switches to the Mayor's house at the end of the passage, but the answer says "majority," so this is still incorrect. Choice *D* is incorrect; we *can* infer that the Lord Mayor is wealthy—he lives in the "Mansion House" and has fifty cooks.

43. D: The passage tells us that the poulterers' and grocers' trades were "a glorious pageant, with which it was next to impossible to believe that such dull principles as bargain and sale had anything to do," which means they set up their shops to be entertaining public spectacles in order to increase sales. Choice *A* is incorrect; although the word "joke" is used, it is meant to be used as a source of amusement rather than something made in poor quality. Choice *B* is incorrect; that they put on a "pageant" is figurative for the public spectacle they made with their shops, not a literal play. Finally, Choice *C* is incorrect, as this is not mentioned anywhere in the passage.

44. B: Proffering means to offer something. Choice *A*, giving away, is incorrect. *Bolstering* means helping or maintaining so Choice *C* is incorrect. Choice *D*, teaching, doesn't make sense in the context of the passage, so this answer is incorrect.

45. A: Choice *A* correctly identifies the argument's primary purpose. The purpose is clearly persuasive, and the focus is on the newspaper's approach to journalism. According to the conclusion, the newspaper isn't currently meeting basic editorial standards, and the journalist wants the newspaper to adopt the best practices described in the argument.

Choice *B* is incorrect. The journalist mentions that the newspaper is currently exposed to charges of bias from all sides, but the argument isn't defending the newspaper. It's calling for a change in editorial policy.

Choice *C* is incorrect. Although the journalist might agree that the newspaper needs to shake up its staff, the primary focus is on the newspaper's approach to journalism.

Choice *D* is incorrect. The journalist touches on the professional responsibilities of a journalist, but it's in the context of the newspaper's failings, which Choice *D* doesn't reference.

46. D: Choice *D* is the best answer choice because the gerunds are all in parallel structure: "Playing baseball, swimming, and dancing." Choice *A* is incorrect because "Play baseball" does not match the parallel structure of the other two gerunds. Choice *B* is incorrect because the answer uses semicolons instead of commas, which is incorrect. Semicolons are used to separate independent clauses. Choice *C* is incorrect because "to swim" and "to dance" eschew parallel structure.

47. B: To convince the audience that judges holding their positions based on good behavior is a practical way to avoid corruption. Choice *A* is incorrect because although he mentions the condition of good behavior as a barrier to despotism, he does not discuss it as a practice in the States. Choice *C* is incorrect because the author does not argue that the audience should vote based on judges' behavior, but rather that good behavior should be the condition for holding their office. Choice *D* is not represented in the passage, so it is incorrect.

48. C: Maritza went to the park to swim with her friends, and when she got there, she realized that they had thrown her a surprise party!" Following the directions carefully will result in this sentence. All the other sentences are close, but Choices *A, B,* and *D* leave at least one of the steps out.

49. D: Glossary. A glossary is a section in a book that provides brief definitions/explanations for words that the reader may not know. Choice *A* is incorrect because a table of contents shows where each section of the book is located. Choice *B* is incorrect because the introduction is usually a chapter that introduces the book about to be read. Choice *C* is incorrect because an index is usually a list of alphabetical references at the end of a book that a reader can look up to get more information.

50. B: Secondary source. This excerpt is considered a secondary source because it actively interprets primary sources. We see direct quotes from the queen, which would be considered a primary source. But since we see those quotes being interpreted and analyzed, the excerpt becomes a secondary source. Choice *C,* tertiary source, is an index of secondary and primary sources, like an encyclopedia or Wikipedia.

51. B: It took two years for the new castle to be built. It states this in the first sentence of the second paragraph. In the third year, we see the Prince planning improvements, and arranging things for the fourth year.

52. C: To impose a certain quality upon. The sentence states that "the impress of his dear hand [has] been stamped everywhere," regarding the quality of his tastes and creations on the house. Choice *A* is one definition of *impress*, but this definition is used more as a verb than a noun: "She impressed us as a songwriter." Choice *B* is incorrect because it is also used as a verb: "He impressed the need for something to be done." Choice *D* is incorrect because it is part of a physical act: "the businessman impressed his mark upon the envelope." The phrase in the passage is meant as figurative, since the workmen did most of the physical labor, not the Prince.

53. B: The passage is describing a cottage, Choice *B.* The name of the building is called "Barton Cottage."

54. D: The middle of the passage says that the narrow passage led through the house into the garden, Choice *D.*

55. A: The passage states that a sitting room was on each side of the entrance.

56. C: There were four bedrooms inside the house, according to the passage.

57. C: The passage is set in the month of September.

58. B: The topic of the passage is about smart assistants, Choice *B.*

59. A: According to the passage, smart devices make your life more efficient, Choice *A.*

60. D: Janie Adams and Ray Gold are both CEO's of big tech companies, Choice *D.*

61. B: According to the passage, a small percentage of our information goes to advertisers.

62. D: While this passage may start out as informative writing, the last paragraph moves it to the persuasive category, because the passage makes a claim: that there is "really no need to worry about safety with smart assistants." We even see the counterargument at the beginning of paragraph two, that "many people believe otherwise" about the harm of smart assistant devices.

Mathematics

Numbers and Operations

Performing Operations on Rational Numbers

Rational numbers are any numbers that can be written as a fraction of integers. Operations to be performed on rational numbers include adding, subtracting, multiplying, and dividing. Essentially, this refers to performing these operations on fractions. Adding and subtracting fractions must be completed by first finding the least common denominator. For example, the problem:

$$\frac{3}{5} + \frac{6}{7}$$

requires that the common multiple be found between 5 and 7. The smallest number that divides evenly by 5 and 7 is 35. For the denominators to become 35, they must be multiplied by 7 and 5 respectively. The fraction $\frac{3}{5}$ can be multiplied by 7 on the top and bottom to yield the fraction $\frac{21}{35}$. The fraction $\frac{6}{7}$ can be multiplied by 5 to yield the fraction $\frac{30}{35}$. Now that the fractions have the same denominator, the numerators can be added. The answer to the addition problem becomes:

$$\frac{3}{5} + \frac{6}{7} = \frac{21}{35} + \frac{30}{35} = \frac{51}{35}$$

The same technique can be used for subtraction of rational numbers. The operations multiplication and division may seem easier to perform because finding common denominators is unnecessary. If the problems reads:

$$\frac{1}{3} \times \frac{4}{5}$$

then the numerators and denominators are multiplied by each other and the answer is found to be $\frac{4}{15}$. For division, the problem must be changed to multiplication before performing operations. The following words can be used to remember to leave, change, and flip before multiplying. If the problems reads:

$$\frac{3}{7} \div \frac{3}{4}$$

then the first fraction is *left* alone, the operation is *changed* to multiplication, and then the last fraction is *flipped*. The problem becomes:

$$\frac{3}{7} \times \frac{4}{3} = \frac{12}{21}$$

Rational numbers can also be negative. When two negative numbers are added, the result is a negative number with an even greater magnitude. When a negative number is added to a positive number, the result depends on the value of each addend. For example:

$$-4 + 8 = 4$$

because the positive number is larger than the negative number. For multiplying two negative numbers, the result is positive. For example:

$$-4 \times -3 = 12$$

where the negatives cancel out and yield a positive answer.

Using Inverse Operations to Solve Problems

Inverse operations can be used to solve problems where there is a missing value. The area for a rectangle may be given, along with the length, but the width may be unknown. This situation can be modeled by the equation Area = Length × Width. The area is 40 square feet and the length is 10 feet. The equation becomes $40 = 10 \times w$. In order to find the w, we recognize that some number multiplied by 10 yields the number 40. The inverse operation to multiplication is division, so the 10 can be divided on both sides of the equation. This operation cancels out the 10 and yields an answer of 4 for the width. The following equation shows the work:

$$40 = 10 \times w$$

$$\frac{40}{10} = \frac{10 \times w}{10}$$

$$4 = w$$

Other inverse operations can be used to solve problems as well. The following equation can be solved for b:

$$b + 4 = 9$$

Because 4 is added to b, it can be subtracted on both sides of the equal sign to cancel out the four and solve for b, as follows:

$$b + 4 - 4 = 9 - 4$$

$$b = 5$$

Whatever operation is used in the equation, the inverse operation can be used and applied to both sides of the equals sign to solve for an unknown value.

Multiplication and Division

A common misconception of multiplication is that it always results in a value greater than the beginning number, or factors. This is not always the case. When working with fractions, **multiplication** may be used to take part of another number. For example:

$$\frac{1}{2} \times \frac{1}{4}$$

can be read as "one-half times one-fourth," or taking one-half of one-fourth. The latter translation makes it easier to understand the concept. Taking half of one-fourth will result in a smaller number that one-fourth. It will result in one-eighth. The same happens with multiplying two-thirds times three-fifths, or:

$$\frac{2}{3} \times \frac{3}{5}$$

The concept of taking two-thirds, which is a part, of three-fifths, means that there will be an even smaller part as the result. Multiplication of these two fractions yields the answer $\frac{6}{15}$, or $\frac{2}{5}$.

In the same way, another misconception is that division always has results smaller than the beginning number or dividend. When working with whole numbers, **division** asks how many times a whole goes into another whole. This result will always be smaller than the dividend, where:

$$6 \div 2 = 3$$

and

$$20 \div 5 = 4$$

When working with fractions, the number of times a part goes into another part depends on the value of each fraction. For example, three-fourths divided by one-fourth, or:

$$\frac{3}{4} \div \frac{1}{4}$$

asks to find how many times $\frac{1}{4}$ will go into $\frac{3}{4}$. Because these have the same denominator, the numerators can be compared as is, without needing to convert the fractions. The result is easily found to be 3 because 1 goes into 3 three times.

Interpreting Remainders in Division Problems
Understanding remainders begins with understanding the division problem. The problem $24 \div 7$ can be read as "twenty-four divided by seven." The problem is asking how many groups of 7 will fit into 24. Counting by seven, the multiples are 7, 14, 21, 28. Twenty-one, which is three groups of 7, is the closest to 24. The difference between 21 and 24 is 3, which is called the remainder. This is a remainder because it is the number that is left out after the three groups of seven are taken from 24. The answer to this division problem can be written as 3 with a remainder 3, or $3\frac{3}{7}$. The fraction $\frac{3}{7}$ can be used because it shows the part of the whole left when the division is complete. Another division problem may have the following numbers: $36 \div 5$. This problem is asking how many groups of 5 will fit evenly into 36. When counting by multiples of 5, the following list is generated: 5, 10, 15, 20, 25, 30, 35, 40. As seen in the list, there are seven groups of five that make 35. To get to the total of 36, there needs to be one additional number. The answer to the division problem would be:

$$36 \div 5 = 7 \, R1$$

or $7\frac{1}{5}$. The fractional part represents the number that cannot make up a whole group of five.

Fractions

Composing and Decomposing Fractions

Fractions are ratios of whole numbers and their negatives. Fractions represent parts of wholes, whether pies, or money, or work. The number on top, or numerator, represents the part, and the bottom number, or denominator, represents the whole. The number $\frac{1}{2}$ represents half of a whole. Other ways to represent one-half are $\frac{2}{4}, \frac{3}{6},$ and $\frac{5}{10}$. These are fractions not written in simplest form, but the numerators are all halves of the denominators. The fraction $\frac{1}{4}$ represents 1 part to a whole of 4 parts. This can be modeled by the quarter's value in relation to the dollar. One quarter is $\frac{1}{4}$ of a dollar. In the same way, 2 quarters make up $\frac{1}{2}$ of a dollar, so 2 fractions of $\frac{1}{4}$ make up a fraction of $\frac{1}{2}$. Three quarters make up three-fourths of a dollar. The three fractions of:

$$\frac{1}{4} + \frac{1}{4} + \frac{1}{4}$$

are equal to $\frac{3}{4}$ of a whole. This illustration can be seen using the bars below divided into one whole, then two halves, then three sections of one-third, then four sections of one-fourth. Based on the size of the fraction, different numbers of each fraction are needed to make up a whole.

Unit Fractions

A **unit fraction** is a fraction where the numerator has a value of one. The fractions one-half, one-third, one-seventh, and one-tenth are all examples of unit fractions. Examples that are not unit fractions include three-fourths, four-fifths, and seven-twelfths. The value of unit fractions changes as the denominator changes, because the numerator is always one. The unit fraction one-half requires two parts to make a whole. The unit fraction one-third requires three parts to make a whole. In the same way, if the unit fraction changes to one-thirteenth, then the number of parts required to make a whole becomes thirteen. An illustration of this is seen in the figure below. As the denominator increases, the size of the parts for each fraction decreases. As the bar goes from one-fourth to one-fifth, the size of the bars decreases, but the size of the denominator increases to five. This pattern continues down the diagram as the bars, or value of the fraction, get smaller, the denominator gets larger:

Comparing Fractions

Comparing fractions requires the use of a common denominator. This necessity can be seen by the two pies below. The first pie has a shaded value of $\frac{2}{10}$ because two pieces are shaded out of the total of ten equal pieces. The second pie has a shaded value of $\frac{2}{7}$ because two pieces are shaded out of a total of seven equal pieces. These two fractions, $\frac{2}{10}$ and $\frac{2}{7}$, have the same numerator, and so a misconception may be that they are equal. By looking at the shaded region in each pie, it is apparent that the fractions are not equal. The numerators are the same, but the denominators are not. Two parts of a whole are not equivalent unless the whole is broken into the same number of parts. To compare the shaded regions, the denominators 7 and 10 must be made equal. The lowest number that the two denominators will both divide evenly into is 70, which is the lowest common denominator. Then the numerators must be converted by multiplying by the opposite denominator. These operations result in the two fractions $\frac{14}{70}$ and $\frac{20}{70}$. Now that these two have the same denominator, the conclusion can be made that $\frac{2}{7}$ represents a larger portion of the pie, as seen in the figure below:

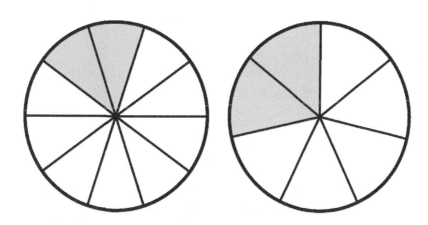

Exponents and Roots

The nth root of a is given as $\sqrt[n]{a}$, which is called a **radical**. Typical values for n are 2 and 3, which represent the square and cube roots. In this form, n represents an integer greater than or equal to 2, and a is a real number. If n is even, a must be nonnegative, and if n is odd, a can be any real number. This radical can be written in exponential form as $a^{\frac{1}{n}}$. Therefore, $\sqrt[4]{15}$ is the same as $15^{\frac{1}{4}}$ and $\sqrt[3]{-5}$ is the same as $(-5)^{\frac{1}{3}}$.

In a similar fashion, the nth root of a can be raised to a power m, which is written as $\left(\sqrt[n]{a}\right)^{m}$. This expression is the same as $\sqrt[n]{a^{m}}$. For example:

$$\sqrt[2]{4^3} = \sqrt[2]{64} = 8 = \left(\sqrt[2]{4}\right)^{3} = 2^3$$

Because:

$$\sqrt[n]{a} = a^{\frac{1}{n}}$$

both sides can be raised to an exponent of m, resulting in:

$$\left(\sqrt[n]{a}\right)^{m} = \sqrt[n]{a^{m}} = a^{\frac{m}{n}}$$

This rule allows:

$$\sqrt[2]{4^3} = \left(\sqrt[2]{4}\right)^{3} = 4^{\frac{3}{2}} = (2^2)^{\frac{3}{2}} = 2^{\frac{6}{2}} = 2^3 = 8$$

Negative exponents can also be incorporated into these rules. Any time an exponent is negative, the base expression must be flipped to the other side of the fraction bar and rewritten with a positive exponent. For instance:

$$2^{-3} = \frac{1}{2^3} = \frac{1}{8}$$

Therefore, two more relationships between radical and exponential expressions are:

$$a^{-\frac{1}{n}} = \frac{1}{\sqrt[n]{a}}$$

$$a^{-\frac{m}{n}} = \frac{1}{\sqrt[n]{a^m}} = \frac{1}{\left(\sqrt[n]{a}\right)^m}$$

Thus:

$$8^{-\frac{1}{3}} = \frac{1}{\sqrt[3]{8}} = \frac{1}{2}$$

All of these relationships are very useful when simplifying complicated radical and exponential expressions. If an expression contains both forms, use one of these rules to change the expression to contain either all radicals or all exponential expressions. This process makes the entire expression much easier to work with, especially if the expressions are contained within equations.

Consider the following example:

$$\sqrt{x} \times \sqrt[4]{x}$$

It is written in radical form; however, it can be simplified into one radical by using exponential expressions first. The expression can be written as

$$x^{\frac{1}{2}} \times x^{\frac{1}{4}}$$

It can be combined into one base by adding the exponents as:

$$x^{\frac{1}{2}+\frac{1}{4}} = x^{\frac{3}{4}}$$

Writing this back in radical form, the result is $\sqrt[4]{x^3}$.

Vectors

A **vector** is something that has both magnitude and direction. A vector may sometimes be represented by a ray that has a length, for its magnitude, and a direction. As the magnitude of the vector increases, the length of the ray changes. The direction of the ray refers to the way that the magnitude is applied. The following vector shows the placement and parts of a vector:

Parts of a Vector

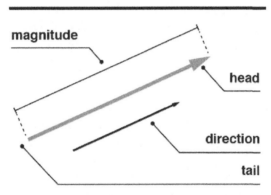

Examples of vectors include force and velocity. **Force** is a vector because applying force requires magnitude, which is the amount of force, and a direction, which is the way a force is applied. **Velocity** is a vector because it has a magnitude, or speed that an object travels, and also the direction that the object is going in. Vectors can be added together by placing the tail of the second at the head of the first. The resulting vector is found by starting at the first tail and ending at the second head. An example of this is show in the following picture:

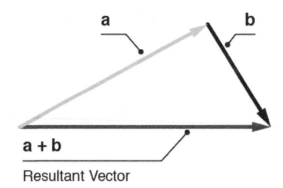

a + b

Resultant Vector

Subtraction can also be done with vectors by adding the inverse of the second vector. The inverse is found by reversing the direction of the vector. Then addition can take place just as described above, but using the inverse instead of the original vector. Scalar multiplication can also be done with vectors. This multiplication changes the magnitude of the vector by the **scalar**, or number. For example, if the length is described as 4, then scalar multiplication is used to multiply by 2, where the vector magnitude becomes 8. The direction of the vector is left unchanged because scalar does not include direction.

Vectors may also be described using coordinates on a plane, such as (5, 2). This vector would start at a point and move to the right 5 and up 2. The two coordinates describe the horizontal and vertical components of the vector. The starting point in relation to the coordinates is the tail, and the ending point is the head.

Matrices

A **matrix** is an arrangement of numbers in rows and columns. Matrices are used to work with vectors and transform them. One example is a system of linear equations. Matrices can represent a system and be used to transform and solve the system. An important connection between scalars, vectors, and matrices is this: scalars are only numbers, vectors are numbers with magnitude and direction, and matrices are an array of numbers in rows and columns. The rows run from left to right and the columns run from top to bottom. When describing the dimensions of a matrix, the number of rows is stated first and the number of columns is stated second. The following matrix has two rows and three columns, referred to as a 2 × 3 matrix:

$$\begin{bmatrix} 3 & 5 & 7 \\ 4 & 2 & 8 \end{bmatrix}$$

A number in a matrix can be found by describing its location. For example, the number in row two, column three is 8. In row one, column two, the number 5 is found.

Operations can be performed on matrices, just as they can on vectors. Scalar multiplication can be performed on matrices and it will change the magnitude, just as with a vector. A scalar multiplication problem using a 2 × 2 matrix looks like the following:

$$3 \times \begin{bmatrix} 4 & 5 \\ 8 & 3 \end{bmatrix}$$

The scalar of 3 is multiplied by each number to form the resulting matrix:

$$\begin{bmatrix} 12 & 15 \\ 24 & 9 \end{bmatrix}$$

Matrices can also be added and subtracted. For these operations to be performed, the matrices must be the same dimensions. Other operations that can be performed to manipulate matrices are multiplication, division, and transposition. **Transposing** a matrix means to switch the rows and columns. If the original matrix has two rows and three columns, then the transposed matrix has three rows and two columns.

Using Structure to Isolate or Identify a Quantity of Interest

When solving equations, it is important to note which quantity must be solved for. This quantity can be referred to as the **quantity of interest**. The goal of solving is to isolate the variable in the equation using logical mathematical steps. The **addition property of equality** states that the same real number can be added to both sides of an equation and equality is maintained. Also, the same real number can be subtracted from both sides of an equation to maintain equality. Second, the **multiplication property of equality** states that the same nonzero real number can multiply both sides of an equation, and still, equality is maintained. Because division is the same as multiplying times a reciprocal, an equation can be divided by the same number on both sides as well.

When solving inequalities, the same ideas are used. However, when multiplying by a negative number on both sides of an inequality, the inequality symbol must be flipped in order to maintain the logic. The same is true when dividing both sides of an inequality by a negative number.

Basically, in order to isolate a quantity of interest in either an equation or inequality, the same thing must be done to both sides of the equals sign, or inequality symbol, to keep everything mathematically correct.

Algebra

Algebraic Expressions

Expressions Versus Equations
An **algebraic expression** is a mathematical phrase that may contain numbers, variables, and mathematical operations. An expression represents a single quantity. For example, $3x + 2$ is an algebraic expression.

An **algebraic equation** is a mathematical sentence with two expressions that are equal to each other. That is, an equation must contain an equals sign, as in:

$$3x + 2 = 17$$

This statement says that the value of the expression on the left side of the equals sign is equivalent to the value of the expression on the right side. In an expression, there are not two sides because there is no equals sign. The equals sign (=) is the difference between an expression and an equation.

To distinguish an expression from an equation, just look for the equals sign.

Example: Determine whether each of these is an expression or an equation.

- $16 + 4x = 9x - 7$ Solution: Equation

- $-27x - 42 + 19y$ Solution: Expression

- $4 = x + 3$ Solution: Equation

Parts of Expressions

A **variable** is a symbol used to represent a number. Letters, like x, y, and z, are often used as variables in algebra.

A **constant** is a number that cannot change its value. For example, 18 is a constant.

A **term** is a constant, variable, or the product of constants and variables. In an expression, terms are separated by $+$ and $-$ signs. Examples of terms are $24x$, -32, and $15xyz$.

Like terms are terms that contain the same variables in the same powers. For example, $6z$ and $-8z$ are like terms, and $9xy$ and $17xy$ are like terms. $12y^3$ and $3y^3$ are also like terms. Lastly, constants, like 23 and 51, are like terms as well.

A **factor** is something that is multiplied by something else. A factor may be a constant, a variable, or a sum of constants or variables.

A **coefficient** is the numerical factor in a term that has a variable. In the term $16x$, the coefficient is 16.

Example: Given the expression:

$$6x - 12y + 18$$

answer the following questions.

1. How many terms are in the expression?

    Solution: 3

2. Name the terms.

    Solution: $6x$, $-12y$, and 18 (Notice that the minus sign preceding the 12 is interpreted to represent negative 12)

3. Name the factors.

    Solution: 6, x, –12, y

4. What are the coefficients in this expression?

    Solution: 6 and –12

5. What is the constant in this expression?

    Solution: 18

Adding and Subtracting Linear Algebraic Expressions

To add and subtract linear algebra expressions, you must combine like terms. **Like terms** are described as those terms that have the same variable with the same exponent. In the following example, the x-terms can be added because the variable is the same and the exponent on the variable of one is also the same. These terms add to be $9x$. The other like terms are called **constants** because they have no variable component. These terms will add to be nine.

Example: Add $(3x - 5) + (6x + 14)$

$3x - 5 + 6x + 14$	Rewrite without parentheses
$3x + 6x - 5 + 14$	Use the commutative property of addition
$9x + 9$	Combine like terms

When subtracting linear expressions, be careful to add the opposite when combining like terms. Do this by distributing -1, which is multiplying each term inside the second parenthesis by negative one. Remember that distributing -1 changes the sign of each term.

Example: Subtract $(17x + 3) - (27x - 8)$

$17x + 3 - 27x + 8$	Use the distributive property
$17x - 27x + 3 + 8$	Use the commutative property of addition
$-10x + 11$	Combine like terms

Example: Simplify by adding or subtracting:

$$(6m + 28z - 9) + (14m + 13) - (-4z + 8m + 12)$$

$6m + 28z - 9 + 14m + 13 + 4z - 8m - 12$	Use the distributive property
$6m + 14m - 8m + 28z + 4z - 9 + 13 - 12$	Use the commutative property of addition
$12m + 32z - 8$	Combine like terms

Using the Distributive Property to Generate Equivalent Linear Algebraic Expressions

The Distributive Property:

$$a(b + c) = ab + ac$$

The **distributive property** is a way of taking a factor and multiplying it through a given expression in parentheses. Each term inside the parentheses is multiplied by the outside factor, eliminating the parentheses. The following example shows how to distribute the number 3 to all the terms inside the parentheses.

Example: Use the distributive property to write an equivalent algebraic expression:

$$3(2x + 7y + 6)$$

$3(2x) + 3(7y) + 3(6)$ Use the distributive property

$6x + 21y + 18$ Simplify

Because $a - b$ can be written $a + (-b)$, the distributive property can be applied in the example below.

Example: Use the distributive property to write an equivalent algebraic expression.

$7(5m - 8)$

$7[5m + (-8)]$ Rewrite subtraction as addition of -8

$7(5m) + 7(-8)$ Use the distributive property

$35m - 56$ Simplify

In the following example, note that the factor of 2 is written to the right of the parentheses but is still distributed as before:

Example: Use the distributive property to write an equivalent algebraic expression:

$(3m + 4x - 10)2$

$(3m)2 + (4x)2 + (-10)2$ Use the distributive property

$6m + 8x - 20$ Simplify

Example: $-(-2m + 6x)$

In this example, the negative sign in front of the parentheses can be interpreted as $-1(-2m + 6x)$

$1(-2m + 6x)$

$-1(-2m) + (-1)(6x)$ Use distributive property

$2m - 6x$ Simplify

Evaluating Simple Algebraic Expressions for Given Values of Variables
To evaluate an algebra expression for a given value of a variable, replace the variable with the given value. Then perform the given operations to simplify the expression.

Example: Evaluate $12 + x$ for $x = 9$

$12 + (9)$ Replace x with the value of 9 as given in the problem. It is a good idea to always use parentheses when substituting this value. This will be particularly important in the following examples.

21 Add

Now see that when x is 9, the value of the given expression is 21.

Example: Evaluate $4x + 7$ for $x = 3$

$\quad\quad\quad 4(3) + 7$ $\quad\quad\quad\quad\quad$ Replace the x in the expression with 3

$\quad\quad\quad 12 + 7$ $\quad\quad\quad\quad\quad\quad$ Multiply (remember order of operations)

$\quad\quad\quad 19$ $\quad\quad\quad\quad\quad\quad\quad\quad$ Add

Therefore, when x is 3, the value of the given expression is 19.

Example: Evaluate $-7m - 3r - 18$ for $m = 2$ and $r = -1$

$\quad\quad\quad -7(2) - 3(-1) - 18$ $\quad\quad$ Replace m with 2 and r with -1

$\quad\quad\quad -14 + 3 - 18$ $\quad\quad\quad\quad$ Multiply

$\quad\quad\quad -29$ $\quad\quad\quad\quad\quad\quad\quad\quad$ Add

So, when m is 2 and r is -1, the value of the given expression is -29.

Using Formulas to Determine Unknown Quantities

Given the formula for the area of a rectangle, $A = lw$, with A = area, l = length, and w = width, the area of a rectangle can be determined, given the length and the width.

For example, if the length of a rectangle is 7 cm and the width is 10 cm, find the area of the rectangle. Just as when evaluating expressions, to solve, replace the variables with the given values. Thus, given $A = lw$, and $l = 7$ and $w = 10$, $A = (7)(10)$, which equals 70. Therefore, the area of the rectangle is 70 cm^2.

Consider an example using the formula for perimeter of a rectangle, which is:

$$P = 2l + 2w$$

where P is perimeter, l is length, and w is width. If the length of a rectangle is 12 inches and the width is 9 inches, find the perimeter.

$\quad\quad$ Solution: $P = 2l + 2w$

$\quad\quad\quad\quad P = 2(12) + 2(9)$ $\quad\quad\quad$ Replace l with 12 and w with 9

$\quad\quad\quad\quad P = 24 + 18$ $\quad\quad\quad\quad\quad$ Use correct order of operations; multiply first

$\quad\quad\quad\quad P = 42$ $\quad\quad\quad\quad\quad\quad\quad$ Add

The perimeter of this rectangle is 42 inches.

Creating, Solving, or Interpreting a Linear Expression or Equation in One Variable

An **equation in one variable** is a mathematical statement where two algebraic expressions in one variable, usually x, are set equal. To solve the equation, the variable must be isolated on one side of the equals sign. The addition and multiplication principles of equality are used to isolate the variable. The **addition principle of equality** states that the same number can be added to or subtracted from both sides of an equation. Because the same value is being used on both sides of the equals sign, equality is maintained. For example, the equation:

$$2x - 3 = 5x$$

is equivalent to both:

$$(2x - 3) + 3 = 5x + 3$$

And

$$(2x - 3) - 5 = 5x - 5$$

This principle can be used to solve the following equation:

$$x + 5 = 4$$

The variable x must be isolated, so to move the 5 from the left side, subtract 5 from both sides of the equals sign. Therefore:

$$x + 5 - 5 = 4 - 5$$

So, the solution is $x = -1$.

This process illustrates the idea of an **additive inverse** because subtracting 5 is the same as adding -5. Basically, add the opposite of the number that must be removed to both sides of the equals sign. The **multiplication principle of equality** states that equality is maintained when a number is either multiplied by both expressions on each side of the equals sign, or when both expressions are divided by the same number. For example:

$$4x = 5$$

is equivalent to both:

$$16x = 20$$

and $x = \frac{5}{4}$. Multiplying both sides times 4 and dividing both sides by 4 maintains equality. Solving the equation:

$$6x - 18 = 5$$

requires the use of both principles. First, apply the addition principle to add 18 to both sides of the equals sign, which results in:

$$6x = 23$$

Then use the multiplication principle to divide both sides by 6, giving the solution:

$$x = \frac{23}{6}$$

Using the multiplication principle in the solving process is the same as involving a multiplicative inverse. A **multiplicative inverse** is a value that, when multiplied by a given number, results in 1. Dividing by 6 is the same as multiplying by $\frac{1}{6}$, which is both the reciprocal and multiplicative inverse of 6.

When solving a linear equation in one variable, checking the answer shows if the solution process was performed correctly. Plug the solution into the variable in the original equation. If the result is a false statement, something was done incorrectly during the solution procedure. Checking the example above gives the following:

$$6 \times \frac{23}{6} - 18 = 23 - 18 = 5$$

Therefore, the solution is correct.

Some equations in one variable involve fractions or the use of the **distributive property**. In either case, the goal is to obtain only one variable term and then use the addition and multiplication principles to isolate that variable. Consider the equation:

$$\frac{2}{3}x = 6$$

To solve for x, multiply each side of the equation by the reciprocal of $\frac{2}{3}$, which is $\frac{3}{2}$. This step results in:

$$\frac{3}{2} \times \frac{2}{3}x = \frac{3}{2} \times 6$$

which simplifies into the solution $x = 9$. Now consider the equation:

$$3(x + 2) - 5x = 4x + 1$$

Use the distributive property to clear the parentheses. Therefore, multiply each term inside the parentheses by 3. This step results in:

$$3x + 6 - 5x = 4x + 1$$

Next, collect like terms on the left-hand side. **Like terms** are terms with the same variable or variables raised to the same exponent(s). Only like terms can be combined through addition or subtraction. After collecting like terms, the equation is:

$$-2x + 6 = 4x + 1$$

Finally, apply the addition and multiplication principles. Add $2x$ to both sides to obtain:

$$6 = 6x + 1$$

Then, subtract 1 from both sides to obtain $5 = 6x$. Finally, divide both sides by 6 to obtain the solution $\frac{5}{6} = x$.

Two other types of solutions can be obtained when solving an equation in one variable. The final result could be that there is either no solution or that the solution set contains all real numbers. Consider the equation:

$$4x = 6x + 5 - 2x$$

First, the like terms can be combined on the right to obtain:

$$4x = 4x + 5$$

Next, subtract $4x$ from both sides. This step results in the false statement $0 = 5$. There is no value that can be plugged into x that will ever make this equation true. Therefore, there is no solution. The solution procedure contained correct steps, but the result of a false statement means that no value satisfies the equation. The symbolic way to denote that no solution exists is $\emptyset$. Next, consider the equation:

$$5x + 4 + 2x = 9 + 7x - 5$$

Combining the like terms on both sides results in:

$$7x + 4 = 7x + 4$$

The left-hand side is exactly the same as the right-hand side. Using the addition principle to move terms, the result is $0 = 0$, which is always true. Therefore, the original equation is true for any number, and the solution set is all real numbers. The symbolic way to denote such a solution set is $\mathbb{R}$, or in interval notation, $(-\infty, \infty)$.

Translating Phrases and Sentences into Expressions, Equations, and Inequalities
When presented with a real-world problem that must be solved, the first step is always to determine what the unknown quantity is that must be solved for. Use a variable, such as x or t, to represent that unknown quantity. Sometimes, there can be two or more unknown quantities. In this case, either choose an additional variable, or if a relationship exists between the unknown quantities, express the other quantities in terms of the original variable. After choosing the variables, form algebraic expressions and/or equations that represent the verbal statement in the problem. The following table shows examples of vocabulary used to represent the different operations:

Addition	Sum, plus, total, increase, more than, combined, in all
Subtraction	Difference, less than, subtract, reduce, decrease, fewer, remain
Multiplication	Product, multiply, times, part of, twice, triple
Division	Quotient, divide, split, each, equal parts, per, average, shared

The combination of operations and variables form both mathematical expression and equations. As mentioned, the difference between expressions and equations are that there is no equals sign in an expression, and that expressions are evaluated to find an unknown quantity, while equations are solved to find an unknown quantity. Also, inequalities can exist within verbal mathematical statements. Instead of a statement of equality, expressions state quantities are *less than*, *less than or equal to*, *greater than*, or *greater than or equal to*. Another type of inequality is when a quantity is said to be *not equal to* another quantity. The symbol used to represent "not equal to" is $\neq$.

The steps for solving inequalities in one variable are the same steps for solving equations in one variable. The addition and multiplication principles are used. However, to maintain a true statement when using the

$<, \le, >$, and $\ge$ symbols, if a negative number is either multiplied by both sides of an inequality or divided from both sides of an inequality, the sign must be flipped. For instance, consider the following inequality:

$$3 - 5x \le 8$$

First, 3 is subtracted from each side to obtain $-5x \le 5$. Then, both sides are divided by -5, while flipping the sign, to obtain $x \ge -1$. Therefore, any real number greater than or equal to -1 satisfies the original inequality.

Linear Relationships Represented by Graphs, Equations, and Tables

Graphs, equations, and tables are three different ways to represent linear relationships. The following graph shows a linear relationship because the relationship between the two variables is constant. As the distance increases by 25 miles, the time lapses by 1 hour. This pattern continues for the rest of the graph. The line represents a constant rate of 25 miles per hour. This graph can also be used to solve problems involving predictions for a future time. After 8 hours of travel, the rate can be used to predict the distance covered. Eight hours of travel at 25 miles per hour covers a distance of 200 miles. The equation at the top of the graph corresponds to this rate also. The same prediction of distance in a given time can be found using the equation.

For a time of 10 hours, the distance would be 250 miles, as the equation yields:

$$d = 25 \times 10 = 250$$

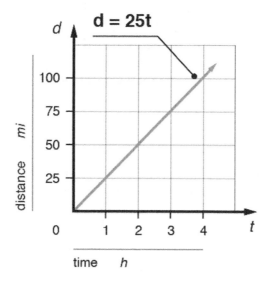

Another representation of a linear relationship can be seen in a table. The first thing to observe from the table is that the y-values increase by the same amount of 3 each time. As the x-values increase by 1, the y-values increase by 3. This pattern shows that the relationship is linear. If this table shows the money earned, y-value, for the hours worked, x-value, then it can be used to predict how much money will be earned for future hours. If 6 hours are worked, then the pay would be $19. For further hours and money to be determined, it would be helpful to have an equation that models this table of values. The equation will show the relationship between x and y. The y-value can each time be determined by multiplying the x-value by 3, then adding 1. The following equation models this relationship:

$$y = 3x + 1$$

Now that there is an equation, any number of hours, x, can be substituted into the equation to find the amount of money earned, y.

y = 3x + 1	
x	y
0	1
1	4
2	7
4	13
5	16

Creating, Solving, or Interpreting a Linear Inequality in One Variable

A **linear equation** *in* x can be written in the form:

$$ax + b = 0$$

A **linear inequality** is very similar, although the equals sign is replaced by an inequality symbol such as $<$, $>$, $\leq$, or $\geq$. In any case, a can never be zero. Some examples of linear inequalities in one variable are:

$$2x + 3 < 0$$

and

$$4x - 2 \leq 0$$

Solving an inequality involves finding the set of numbers that when plugged into the variable, make the inequality a true statement. These numbers are known as the **solution set** of the inequality. To solve an inequality, use the same properties that are necessary in solving equations. First, add or subtract variable terms and/or constants to obtain all variable terms on one side of the equals sign and all constant terms on the other side.

Then, either multiply both sides times the same number, or divide both sides by the same number, to obtain an inequality that gives the solution set. When multiplying times, or dividing by, a negative number in an inequality, change the direction of the inequality symbol. The solution set can be graphed on a number line. Consider the linear inequality:

$$-2x - 5 > x + 6$$

First, add 5 to both sides and subtract $-x$ off of both sides to obtain:

$$-3x > 11$$

Then, divide both sides by -3, making sure to change the direction of the inequality symbol. These steps result in the solution:

$$x < -\frac{11}{3}$$

Therefore, any number less than $-\frac{11}{3}$ satisfies this inequality.

Solving Inequalities

Inequalities can be solved in a similar method as equations. Basically, the goal is to isolate the variable, and this process can be completed by adding numbers onto both sides, subtracting numbers off of both sides, multiplying numbers onto both sides, and dividing numbers off of both sides of the inequality. Basically, if something is done to one side, it has to be done to the other side, just like when solving equations. However, there is one important difference, and that difference occurs when multiplying times negative numbers and dividing by negative numbers. If either one of these steps must be performed in the solution process, the inequality symbol must be reversed. Consider the following inequality:

$$2 - 3x < 11$$

The goal is to isolate the variable x, so first subtract 2 off both sides to obtain:

$$-3x < 9$$

Then divide both sides by -3, making sure to flip the sign. This results in $x > -3$, which is the solution set. This solution set means that all numbers greater than -3 satisfy the original inequality, and therefore any number larger than -3 is a solution. In **set-builder notation**, this set can be written as:

$$\{x | x > -3\}$$

which is read "all x values such that x is greater than −3." In addition to the inequality form of the solution, solutions of inequalities can be expressed by using both a **number line** and **interval notation**. Here is a chart that highlights all three types of expressing the solutions:

Interval Notation	Number Line Sketch	Set-builder Notation
(a , b)	∘———∘ a b	{ x | a < x < b}
$(a , b]$	∘———● a b	{ x | a < x ≤ b}
$[a , b)$	●———∘ a b	{ x | a ≤ x < b}
$[a , b]$	●———● a b	{ x | a ≤ x ≤ b}
(a , ∞)	∘——— a	{ x | x > a}
$(-\infty , b)$	———∘ b	{ x | x < b}
$[a , \infty)$	●——— a	{ x | x ≥ a}
$(-\infty , b]$	———● b	{ x | x ≤ b}
$(-\infty, \infty)$	———————	$\mathbb{R}$

Algebraically Solving Linear Equations or Inequalities in One Variable
A **linear equation in one variable** can be solved using the following steps:

1. Simplify the algebraic expressions on both sides of the equals sign by removing all parentheses, using the distributive property, and then collecting all like terms.

2. Collect all variable terms on one side of the equals sign and all constant terms on the other side by adding the same quantity to both sides of the equals sign, or by subtracting the same quantity from both sides of the equals sign.

3. Isolate the variable by either dividing both sides of the equation by the same number, or by multiplying both sides by the same number.

4. Check the answer.

The only difference between solving linear inequalities versus equations is that when multiplying by a negative number or dividing by a negative number, the direction of the inequality symbol must be reversed.

If an equation contains multiple fractions, it might make sense to clear the equation of fractions first by multiplying all terms by the least common denominator. Also, if an equation contains several decimals, it might make sense to clear the decimals as well by multiplying times a factor of 10. If the equation has decimals in the hundredths place, multiply every term in the equation by 100.

188

Multistep one-variable equations involve the use of one variable in an equation with many operations. For example, the equation $2x + 4 = 10$ involves one variable, x, and multiple steps to solve for the value of x. The first step is to move the 4 to the opposite side of the equation by subtracting 4. The next step is to divide by 2. The final answer yields a value of 3 for the variable x. The steps for this process are shown below:

$$2x + 4 = 10$$

$$2x = 6 \qquad \text{Subtract 4 on both sides}$$

$$x = 3 \qquad \text{Divide by 2 on both sides}$$

When the result is found, the value of the variable must be interpreted. For this problem, a value of 3 can be understood as the amount that can be doubled and then increased by 4 to yield a value of 10.

Inequalities can also be interpreted in much the same way. The following inequality can be solved to find the value of b:

$$\frac{b}{7} - 8 \geq 7$$

This inequality models the amount of money a group of friends earned for cleaning up a neighbor's yard, b. There were 7 friends, so the money had to be split seven times. Then $8 was taken away from each friend to pay for materials they bought to help clean the yard. All these things needed to be less than or equal to seven for the friends to each receive at least $7. The first step is to add 8 to both sides of the inequality. Then the 7 can be multiplied on each side. The resulting inequality is:

$$b \geq 105$$

Because the answer is not only an equals sign, the value for b is not a single number. In this problem, the answer communicates that the value of b must be greater than or equal to $105 in order for each friend to make at least $7 for their work. The number for b, what they are paid, can be more than 105 because that would mean they earned more money. They do not want it to be less than 105 because their profit will drop below $7 per piece.

Solving Quadratic Equations

A **quadratic equation** in standard form:

$$ax^2 + bx + c = 0$$

can have either two solutions, one solution, or two complex solutions (no real solutions). This is determined using the determinant:

$$b^2 - 4ac$$

If the determinant is positive, there are two real solutions. If the determinant is negative, there are no real solutions. If the determinant is equal to zero, there is one real solution. For example, given the quadratic equation:

$$4x^2 - 2x + 1 = 0$$

its determinant is:

$$(-2)^2 - 4(4)(1) = 4 - 16 = -12$$

so it has two complex solutions, meaning no real solutions.

There are quite a few ways to solve a quadratic equation. The first is by factoring. If the equation is in standard form and the polynomial can be factored, set each factor equal to 0, and solve using the **Principle of Zero Products**. For example:

$$x^2 - 4x + 3 = (x - 3)(x - 1)$$

Therefore, the solutions of:

$$x^2 - 4x + 3 = 0$$

are those that satisfy both:

$$x - 3 = 0$$

and

$$x - 1 = 0$$

or

$$x = 3$$

and

$$x = 1$$

This is the simplest method to solve quadratic equations; however, not all polynomials inside the quadratic equations can be factored.

Another method is **completing the square**. The polynomial:

$$x^2 + 10x - 9$$

cannot be factored, so the next option is to complete the square in the equation:

$$x^2 + 10x - 9 = 0$$

to find its solutions. The first step is to add 9 to both sides, moving the constant over to the right side, resulting in:

$$x^2 + 10x = 9$$

Then the coefficient of x is divided by 2 and squared. This result is then added to both sides of the equation. In this example:

$$\left(\frac{10}{2}\right)^2 = 25$$

is added to both sides of the equation to obtain:

$$x^2 + 10x + 25 = 9 + 25 = 34$$

The left-hand side can then be factored into:

$$(x + 5)^2 = 34$$

Solving for x then involves taking the square root of both sides and subtracting 5. This leads to the two solutions:

$$x = \pm\sqrt{34} - 5$$

The third method is the **quadratic formula**. Given a quadratic equation in standard form, $ax^2 + bx + c = 0$, its solutions always can be found using the formula:

$$x = \frac{-b \pm \sqrt{b^2 - 4ac}}{2a}$$

Functions

A **relation** is any set of ordered pairs (x, y). The first set of points, known as the x-coordinates, make up the domain of the relation. The second set of points, known as the y-coordinates, make up the range of the relation. A relation in which every member of the domain corresponds to only one member of the range is known as a **function.** A function cannot have a member of the domain corresponding to two members of the range. Functions are most often given in terms of equations instead of ordered pairs. For instance, here is an equation of a line: $y = 2x + 4$. In function notation, this can be written as:

$$f(x) = 2x + 4$$

The expression $f(x)$ is read "f of x" and it shows that the inputs, the x-values, get plugged into the function and the output is $y = f(x)$. The set of all inputs are in the domain and the set of all outputs are in the range.

The x-values are known as the **independent variables** of the function and the y-values are known as the **dependent variables** of the function. The y-values depend on the x-values. For instance, if $x = 2$ is plugged into the function shown above, the y-value depends on that input.

$$f(2) = 2 \times 2 + 4 = 8.$$

Therefore, $f(2) = 8$, which is the same as writing the ordered pair (2, 8). To graph a function, graph it in equation form and plot ordered pairs.

Due to the definition of a function, the graph of a function cannot have two of the same x-components paired to different y-component. For example, the ordered pairs (3, 4) and (3, -1) cannot be in a valid function. Therefore, all graphs of functions pass the **vertical line test**. If any vertical line intersects a graph in more than one place, the graph is not that of a function. For instance, the graph of a circle is not a function because one can draw a vertical line through a circle and the line would intersect the circle twice. Common functions include lines and polynomials, and they all pass the vertical line test.

Even and Odd Functions
A function is considered *even* when:

$$f(x) = f(-x)$$

for all values of x. This relationship means that the graph of an even function is perfectly symmetrical about the y-axis. In other words, the graph is reflected over the y-axis. The term "even" describes these functions because functions like x^2, x^4, x^6, and so on display this characteristic. That said, there are some functions that are even that don't involve x raised to an even exponent. For example, the graph of:

$$f(x) = cos(x)$$

is also symmetrical over the y-axis. There are also functions that *do* have an even exponent that are *not* even functions; although:

$$f(x) = x^2 + 3$$

is an even function:

$$f(x) = (x + 3)^2$$

is not because:

$$f(x) \neq f(-x)$$

for all values of x.

A function is considered odd when:

$$-f(x) = f(-x)$$

for all values of x. This relationship means that the graph of an odd function is perfectly symmetrical about the origin. The term "odd" describes these functions because functions like x, x^3, x^5, x^7, and so on display this characteristic. As with even functions, exceptions to these exponential values exist. For example, the graph of:

$$f(x) = sin(x)$$

also has origin symmetry. There are also functions that *do* have an odd exponent that are *not* odd functions; while:

$$f(x) = x^3 - 3x$$

is an odd function:

$$f(x) = x^3 - 1$$

is not because $-f(x) \neq f(-x)$ for all values of x.

The majority of functions do not display these unique relationships and are thus neither even nor odd. Essentially, to algebraically determine if a function is even, odd, or neither, various values need to be plugged in for $f(x)$ and $f(-x)$. If all values of x yield the same output for $f(x)$ and $f(-x)$, the function is even. If all values of x yield the same output for $-f(x)$ and $f(-x)$, the function is odd. Any other situation indicates the function is neither even nor odd. To graphically determine if a function is even, odd, or neither, symmetry needs to be evaluated. If the function is perfectly symmetrical across the y-axis, it is an even function. If it is symmetrical about the origin, the function is odd. Any other situation indicates that the function is neither even nor odd.

It should be noted that there is one function that is both even *and* odd: $f(x) = 0$.

Building a Linear Function that Models a Linear Relationship Between Two Quantities
A linear function that models a linear relationship between two quantities is of the form $y = mx + b$, or in function form $f(x) = mx + b$. In a linear function, the value of y depends on the value of x, and y increases or decreases at a constant rate as x increases. Therefore, the independent variable is x, and the dependent variable is y. The graph of a linear function is a line, and the constant rate can be seen by looking at the steepness, or slope, of the line. If the line increases from left to right, the slope is positive. If the line slopes downward from left to right, the slope is negative. In the function, m represents slope. Each point on the line is an **ordered pair** (x, y), where x represents the x-coordinate of the point and y represents the y-coordinate of the point.

The point where $x = 0$ is known as the y-intercept, and it is the place where the line crosses the y-axis. If $x = 0$ is plugged into $f(x) = mx + b$, the result is $f(0) = b$, so therefore, the point $(0, b)$ is the y-intercept of the line. The derivative of a linear function is its slope.

Consider the following situation. A taxicab driver charges a flat fee of $2 per ride and $3 a mile. This statement can be modeled by the function $f(x) = 3x + 2$ where x represents the number of miles and $f(x) = y$ represents the total cost of the ride. The total cost increases at a constant rate of $2 per mile, and that is why this situation is a linear relationship. The slope $m = 3$ is equivalent to this rate of change. The flat fee of $2 is the y-intercept. It is the place where the graph crosses the x-axis, and it represents the cost when $x = 0$, or when no miles have been traveled in the cab. The y-intercept in this situation represents the flat fee.

Interpreting the Variables and Constants in Expressions for Linear Functions within the Context Presented
A linear function of the form $f(x) = mx + b$ has two important quantities: m and b. The quantity m represents the slope of the line, and the quantity b represents the y-intercept of the line. When the function represents an actual real-life situation, or mathematical model, these two quantities are very meaningful. The slope, m, represents the rate of change, or the amount y increases or decreases given an increase in x. If m is positive, the rate of change is positive, and if m is negative, the rate of change is negative. The y-intercept, b, represents the amount of the quantity y when x is zero. In many applications, if the x-variable is never a negative quantity, the y-intercept represents the initial amount of the quantity y. Often the x-variable represents time, so it makes sense that the x-variable is never negative.

Consider the following example. These two equations represent the cost, C, of t-shirts, x, at two different printing companies:

$$C(x) = 7x$$

$$C(x) = 5x + 25$$

The first equation represents a scenario that shows the cost per t-shirt is $7. In this equation, x varies directly with y. There is no y-intercept, which means that there is no initial cost for using that printing company. The rate of change is 7, which is price per shirt. The second equation represents a scenario that has both an initial cost and a cost per t-shirt. The slope 5 shows that each shirt is $5. The y-intercept 25 shows that there is an initial cost of using that company. Therefore, it makes sense to use the first company at $7 a shirt when only purchasing a small number of t-shirts. However, any large orders would be cheaper by going with the second company because eventually that initial cost will be negligible.

Polynomials

A **polynomial function** is a function containing a polynomial expression, which is an expression containing constants and variables combined using the four mathematical operations. The degree of a polynomial in one variable is the largest exponent seen on any variable in the expression. Typical polynomial functions are **quartic**, with a degree of 4, **cubic**, with a degree of 3, and **quadratic**, with a degree of 2. Note that the exponents on the variables can only be nonnegative integers. The domain of any polynomial function is all real numbers because any number plugged into a polynomial expression grants a real number output. An example of a quartic polynomial equation is:

$$y = x^4 + 3x^3 - 2x + 1$$

The zeros of a polynomial function are the points where its graph crosses the y-axis. In order to find the number of real zeros of a polynomial function, **Descartes' Rule of Sign** can be used. The number of possible positive real zeros is equal to the number of sign changes in the coefficients of the terms in the polynomial.

If there is only one sign change, there is only one positive real zero. In the example above, the signs of the coefficients are positive, positive, negative, and positive. Therefore, the sign changes two times and therefore, there are at most two positive real zeros. The number of possible negative real zeros is equal to the number of sign changes in the coefficients when plugging $-x$ into the equation. Again, if there is only one sign change, there is only one negative real zero. The polynomial result when plugging $-x$ into the equation is:

$$y^4 - 3x^3 + 2x + 1$$

The sign changes two times, so there are at most two negative real zeros. Another polynomial equation this rule can be applied to is:

$$y = x^3 + 2x - x - 5$$

There is only one sign change in the terms of the polynomial, so there is exactly one real zero. When plugging $-x$ into the equation, the polynomial result is:

$$-x^3 - 2x - x - 5$$

There are no sign changes in this polynomial, so there are no possible negative zeros.

Adding, Subtracting, and Multiplying Polynomial Equations

When working with polynomials, **like terms** are terms that contain exactly the same variables with the same powers. For example, $x^4 y^5$ and $9x^4 y^5$ are like terms. The coefficients are different, but the same variables are raised to the same powers. When adding polynomials, only terms that are like can be added. When adding two like terms, just add the coefficients and leave the variables alone. This process uses the distributive property. For example:

$$x^4 y^5 + 9x^4 y^5 = (1 + 9)x^4 y^5$$

$$10x^4 y^5$$

Therefore, when adding two polynomials, simply add the like terms together. Unlike terms cannot be combined.

Subtracting polynomials involves adding the opposite of the polynomial being subtracted. Basically, the sign of each term in the polynomial being subtracted is changed, and then the like terms are combined because it is now an addition problem. For example, consider the following:

$$6x^2 - 4x + 2 - (4x^2 - 8x + 1).$$

Add the opposite of the second polynomial to obtain:

$$6x^2 - 4x + 2 + (-4x^2 + 8x - 1)$$

Then, collect like terms to obtain:

$$2x^2 + 4x + 1$$

Multiplying polynomials involves using the product rule for exponents that:

$$b^m b^n = b^{m+n}$$

Basically, when multiplying expressions with the same base, just add the exponents. Multiplying a monomial by a monomial involves multiplying the coefficients together and then multiplying the variables together using the product rule for exponents. For instance:

$$8x^2 y \times 4x^4 y^2 = 32x^6 y^3$$

When multiplying a monomial by a polynomial that is not a monomial, use the distributive property to multiply each term of the polynomial times the monomial. For example:

$$3x(x^2 + 3x - 4) = 3x^3 + 9x^2 - 12x$$

Finally, multiplying two polynomials when neither one is a monomial involves multiplying each term of the first polynomial times each term of the second polynomial. There are some shortcuts, given certain scenarios. For instance, a binomial times a binomial can be found by using the **FOIL** (Firsts, Outers, Inners, Lasts) method shown here.

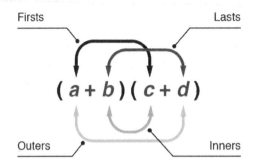

Finding the product of a sum and difference of the same two terms is simple because if it was to be foiled out, the outer and inner terms would cancel out. For instance:

$$(x + y)(x - y) = x^2 + xy - xy - y^2$$

Finally, the square of a binomial can be found using the following formula:

$$(a \pm b)^2 = a^2 \pm 2ab + b^2$$

The Relationship Between Zeros and Factors of Polynomials

A **polynomial** is a mathematical expression containing the sum and difference of one or more terms that are constants multiplied times variables raised to positive powers. A **polynomial equation** is a polynomial set equal to another polynomial, or in standard form, a polynomial is set equal to zero. A **polynomial function** is a polynomial set equal to y. For instance, $x^2 + 2x - 8$ is a polynomial, $x^2 + 2x - 8 = 0$ is a polynomial equation, and $y = x^2 + 2x - 8$ is the corresponding polynomial function. To solve a polynomial equation, the x-values in which the graph of the corresponding polynomial function crosses the x-axis are sought.

These coordinates are known as the **zeros** of the polynomial function, because they are the coordinates in which the y-coordinates are zero. One way to find the zeros of a polynomial is to find its factors, then set each individual factor equal to zero, and solve each equation to find the zeros. A **factor** is a linear expression, and to completely factor a polynomial, the polynomial must be rewritten as a product of individual linear factors. The polynomial listed above can be factored as:

$$(x + 4)(x - 2)$$

Setting each factor equal to zero results in the zeros $x = -4$ and $x = 2$.

Here is the graph of the zeros of the polynomial:

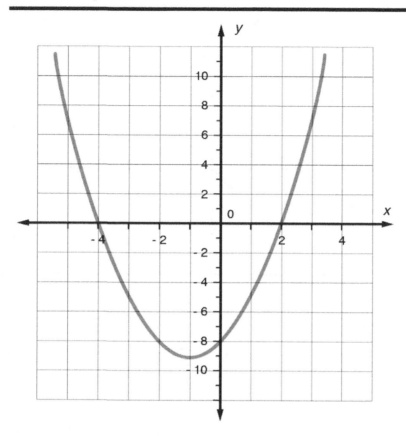

The Graph of the Zeros of x² + 2x - 8 = 0

Systems of Equations

Creating, Solving, or Interpreting Systems of Two Linear Equations in Two Variables
An example of a **system of two linear equations in two variables** is the following:

$$2x + 5y = 8$$

$$5x + 48y = 9$$

A solution to a **system of two linear equations** is an ordered pair that satisfies both the equations in the system. A system can have one solution, no solution, or infinitely many solutions. The solution can be found through a graphing technique. The solution of a system of equations is actually equal to the point of intersection of both lines. If the lines intersect at one point, there is one solution and the system is said to be **consistent**. However, if the two lines are **parallel**, they will never intersect and there is no solution. In this case, the system is said to be **inconsistent**. Third, if the two lines are actually the same line, there are infinitely many solutions and the solution set is equal to the entire line. The lines are **dependent**. Here is a summary of the three cases:

197

Solving Systems by Graphing

Consistent	**Inconsistent**	**Dependent**
One solution	No solution	Infinite number of solutions
Lines intersect	*Lines are parallel*	*Coincide: same line*

Consider the following system of equations:

$$y + x = 3$$

$$y - x = 1$$

To find the solution graphically, graph both lines on the same xy-plane. Graph each line using either a table of ordered pairs, the x- and y-intercepts, or slope and the y-intercept. Then, locate the point of intersection.

The graph is shown here:

The System of Equations $\begin{cases} y + x = 3 \\ y - x = 1 \end{cases}$

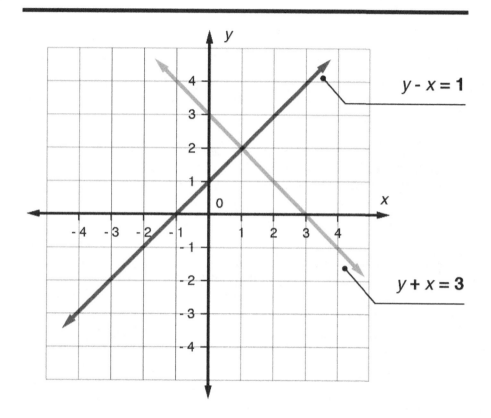

It can be seen that the point of intersection is the ordered pair (1, 2). This solution can be checked by plugging it back into both original equations to make sure it results in true statements. This process results in:

$$2 + 1 = 3$$

$$2 - 1 = 1$$

Both are true equations; therefore, the point of intersection is truly the solution.

The following system has no solution:

$$y = 4x + 1$$

$$y = 4x - 1$$

Both lines have the same slope and different y-intercepts; therefore, they are parallel. This means that they run alongside each other and never intersect.

Finally, the following solution has infinitely many solutions:

$$2x - 7y = 12$$

$$4x - 14y = 24$$

Note that the second equation is equal to the first equation multiplied by 2. Therefore, they are the same line. The solution set can be written in set notation as $\{(x, y) | 2x - 7y = 12\}$, which represents the entire line.

Creating, Solving, or Interpreting Systems of Linear Inequalities in Two Variables
A **system of linear inequalities in two variables** consists of two inequalities in two variables, x and y. For example, the following is a system of linear inequalities in two variables:

$$\begin{cases} 4x + 2y < 1 \\ 2x - y \leq 0 \end{cases}$$

The curly brace on the left side shows that the two inequalities are grouped together. A solution of a single inequality in two variables is an ordered pair that satisfies the inequality. For example, (1, 3) is a solution of the linear inequality $y \geq x + 1$ because when plugged in, it results in a true statement. The graph of an inequality in two variables consists of all ordered pairs that make the solution true. Therefore, the entire solution set of a single inequality contains many ordered pairs, and the set can be graphed by using a half plane. A **half plane** consists of the set of all points on one side of a line. If the inequality consists of $>$ or $<$, the line is dashed because no solutions actually exist on the line shown. If the inequality consists of $\geq$ or $\leq$, the line is solid, and solutions are on the line shown. To graph a linear inequality, graph the corresponding equation found by replacing the inequality symbol with an equals sign. Then pick a test point that exists on either side of the line. If that point results in a true statement when plugged into the original inequality, shade in the side containing the test point. If it results in a false statement, shade in the opposite side.

Solving a system of linear inequalities must be done graphically. Follow the process as described above for both given inequalities. The solution set to the entire system is the region that is in common to every graph in the system.

For example, here is the solution to the following system:

$$\begin{cases} y \geq 3 - x \\ y \leq -3 - x \end{cases}$$

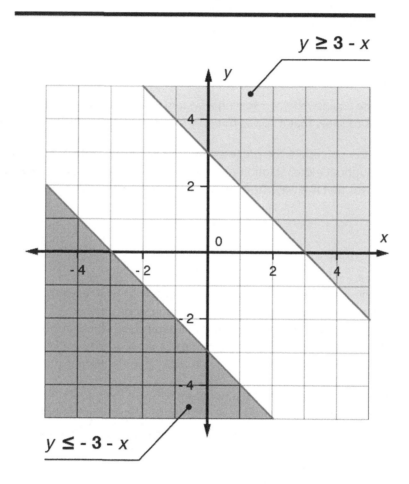

The solution to $\begin{cases} y \geq 3 - x \\ y \leq -3 - x \end{cases}$

$y \geq 3 - x$

$y \leq -3 - x$

Note that there is no region in common, so this system has no solution.

Algebraically Solving Systems of Two Linear Equations in Two Variables
There are two algebraic methods to finding solutions. The first is **substitution**. This process is better suited for systems when one of the equations is already solved for one variable, or when solving for one variable is easy to do. The equation that is already solved for is substituted into the other equation for that variable, and this process results in a linear equation in one variable. This equation can be solved for the given variable, and then that solution can be plugged into one of the original equations, which can then be solved for the other variable. This last step is known as **back-substitution** and the end result is an ordered pair.

A system that is best suited for substitution is the following:

$$y = 4x + 2$$

$$2x + 3y = 9$$

The other method is known as **elimination**, or the **addition method**. This is better suited when the equations are in standard form $Ax + By = C$. The goal in this method is to multiply one or both equations times numbers that result in opposite coefficients. Then, add the equations together to obtain an equation in one variable. Solve for the given variable, then take that value and back-substitute to obtain the other part of the ordered pair solution.

A system that is best suited for elimination is the following:

$$2x + 3y = 8$$

$$4x - 2y = 10$$

Note that in order to check an answer when solving a system of equations, the solution must be checked in both original equations to show that it solves not only one of the equations, but both of them.

If throughout either solution procedure the process results in an untrue statement, there is no solution to the system. Finally, if throughout either solution procedure the process results in the variables dropping out, which gives a statement that is always true, there are infinitely many solutions.

Geometry

Lines

Two lines are **parallel** if they never intersect. Given the equation of two lines, they are parallel if they have the same slope and different y-intercepts. If they had the same slope and same y-intercept, they would be the same line. Therefore, in order to show two lines are parallel, put them in slope-intercept form, $y = mx + b$, to find m and b. The two lines $y = 2x + 6$ and $4x - 2y = 6$ are parallel. The second line in slope intercept is $y = 2x - 3$. Both lines have the same slope, 2, and different y-intercepts.

Two lines are **perpendicular** if they intersect at a right angle. Given the equation of two lines, they are perpendicular if their slopes are negative reciprocals. Therefore, the product of both slopes is equal to -1. For example, the lines $y = 4x + 1$ and $y = -\frac{1}{4}x + 1$ are perpendicular because their slopes are negative reciprocals. The product of 4 and $-\frac{1}{4}$ is -1.

Angle Measurement

In geometry, a **line** connects two points and extends indefinitely in both directions beyond each point. If the length is finite, it is known as a **line segment** and has two end points at either end. A **ray** is a straight path that has one end point and extends indefinitely in the other direction. When lines are extended indefinitely, an arrow is used instead of a point. An angle is formed when two rays begin at the same end point and both extend indefinitely in different directions. The common end point is called a **vertex**. Adjacent angles are formed from two angles using one shared ray. They are two side-by-side angles that also share an end point. Angles are measured in degrees, and their measurement assesses rotation. A full rotation equals 360 degrees and represents a circle. Half of a rotation equals 180 degrees and represents a half-circle. Ninety degrees represents a quarter-circle, which is known as a **right angle**. Any angle less than 90 degrees is called an **acute angle**, and any angle greater than 90 degrees is called an **obtuse angle**. Angle measurement is **additive**, meaning if an angle is broken up into two non-overlapping angles, the total measurement of the larger angle is the sum of the two smaller angles. A protractor can be used to measure an angle. Here is a picture of a protractor measuring a right angle:

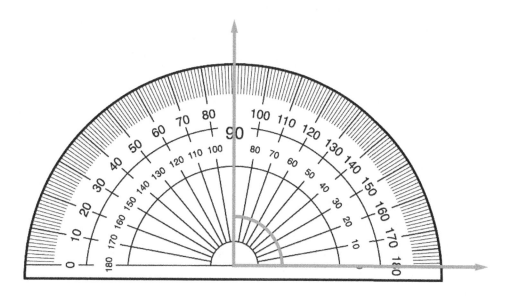

Circles

A **circle** is defined to be the set of all points the same distance, known as **radius** r, from a single point C, known as the **center**. A circle measures 360 degrees. The radius is the length from the center to any point on its edge. Multiply the radius times 2, to obtain the diameter, which is the distance from any two points on the circle that goes through the center. An arc is defined to be all points between any two points on the edge of a circle. A sector can be built from an arc and two corresponding radii. Finally, a central angle is the angle formed by the intersection of those two radii within a sector. Here is a picture that highlights all of these definitions:

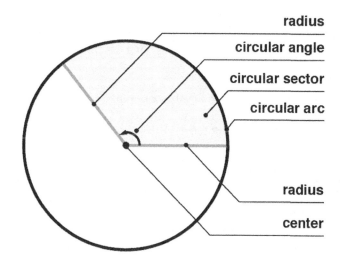

Similar to perimeter, the **circumference** of a circle is equal to the total distance around the outside. The formula for circumference is $C = 2\pi r$, which is the same as $C = \pi d$. The formula for arc length is:

$$2\pi r \, \frac{\text{central angle of arc meaasurement}}{360}$$

The units for both circumference and arc length are linear units, like inches or centimeters, since they are a measurement of length. The formula for area of a circle is $A = \pi r^2$, and the formula for area of a sector is:

$$\pi r^2 \, \frac{\text{central angle of arc meaasurement}}{360}$$

Both units of area are square units.

Triangles

Within a triangle, the measurement of all three angles adds up to 180 degrees. There are three types of special triangles. An **equilateral triangle** has three equal sides and three equal angles, which are each 60 degrees. An **isosceles triangle** has two equal sides, and therefore the measurement of the angles opposite the two equal sides are equal as well. Finally, a **right triangle** has one right angle. The side across from the right angle is known as the **hypotenuse**, and the other two sides are known as the **legs**. There is a special type of right triangle known as a **30-60-90 triangle** because the three angles in the triangle measure 30, 60, and 90 degrees respectively.

The legs of this type of triangle have set relationships due to their corresponding angles. Because the sum of the measures of three angles in any triangle is 180 degrees, if only two angles are known inside a triangle, the third can be found by subtracting the sum of the two known quantities from 180. Two angles whose sum is equal to 90 degrees are known as **complementary angles**. For example, angles measuring 72 and 18 degrees are complementary, and each angle is a complement of the other. Finally, two angles whose sum is equal to 180 degrees are known as **supplementary angles**. To find the supplement of an angle, subtract the given angle from 180 degrees. For example, the supplement of an angle that is 50 degrees is:

$$180 - 50 = 130 \text{ degrees}$$

These terms involving angles can be seen in many types of word problems. For example, consider the following problem: The measure of an angle is 60 degrees less than two times the measure of its complement. What is the angle's measure? To solve this, let x be the unknown angle. Therefore, its complement is $90 - x$. The problem gives that:

$$x = 2(90 - x) - 60$$

To solve for x, distribute the 2, and collect like terms. This process results in:

$$x = 120 - 2x$$

Then, use the addition property to add $2x$ to both sides to obtain $3x = 120$. Finally, use the multiplication properties of equality to divide both sides by 3 to get $x = 40$. Therefore, the angle measures 40 degrees. Also, its complement measures 50 degrees.

Pythagorean Theorem

The **Pythagorean theorem** states that given a right triangle, the sum of the squares of the two legs equals the square of the hypotenuse. For example, consider the following right triangle:

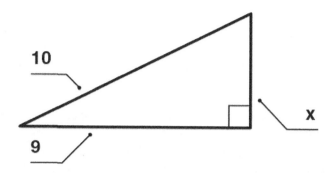

The missing side, x, can be found using the Pythagorean theorem. Since $9^2 + x^2 = 10^2$, $81 + x^2 = 100$, which gives $x^2 = 19$. To solve for x, take the square root of both sides. Therefore, $x = \sqrt{19} = 4.36$, which has been rounded to two decimal places.

Quadrilaterals

A **quadrilateral** is any four-sided polygon, such as a square, rectangle, parallelogram, or trapezoid. Basically, a quadrilateral is a closed shape with four sides. A **parallelogram** is a specific type of quadrilateral that has two sets of parallel lines having the same length. A **trapezoid** is a quadrilateral having only one set of parallel sides. A **rectangle** is a parallelogram that has four right angles and two pairs of equal sides, the length and width or the base and height. A **rhombus** is a parallelogram with two acute angles, two obtuse angles, and four equal sides. The acute angles are of equal measure, and the obtuse angles are of equal measure. Finally, a **square** is a rhombus consisting of four right angles with all sides equal in length. It is important to note that some of these shapes share common attributes. For instance, all four-sided shapes are quadrilaterals. All squares are rectangles, but not all rectangles are squares.

Polygons

A **polygon** is any closed figure consisting of three or more line segments. A three-sided polygon is known as a **triangle**, a four-sided polygon is known as a **quadrilateral**, a five-sided polygon is known as a **pentagon**, a six-sided polygon is known as a **hexagon**, an eight-sided polygon is known as an **octagon**, and a ten-sided polygon is known as a **decagon**. In order to calculate the perimeter of any polygon, just add up the length of all of its sides. For a polygon with n sides, the sum of all angles is determined by the following formula:

$$(n - 2) \times 180°$$

For example, a pentagon has five sides, so the sum of all of its angles is:

$$(5 - 2) \times 180° = 540°$$

A **regular polygon** is defined as one in which all sides and angles are of equal measure. For example, a regular three-sided polygon is an equilateral triangle, and a regular four-sided polygon is a square. In a regular polygon, the measure of each angle is found by dividing the sum of all angles by the number of angles. For instance, a regular pentagon contains five angles, and each has measure $\frac{540}{5} = 108$ degrees.

Congruent and Similar Figures

Two figures are **congruent** if they have the same shape and same size, meaning the same angle measurements and equal side lengths. Two figures are **similar** if they have the same angle measurement but not side lengths. In other words, angles are congruent in similar triangles. Therefore, proving figures are congruent involves showing all angles and sides are the same, and proving figures are similar just involves proving the angles are the same.

If two triangles have two corresponding pairs of angles, the triangles are similar because two angles of equal measure implies equality of the third, since all three add up to 180 degrees. The criteria for triangles to be similar also involve proportionality of side lengths. Corresponding sides of two triangles are sides that are in the same location in the two different shapes. In similar triangles, corresponding side lengths need to be a constant multiple of each other, meaning the ratios of the lengths of those corresponding lengths are all equal.

This is highlighted in the following example:

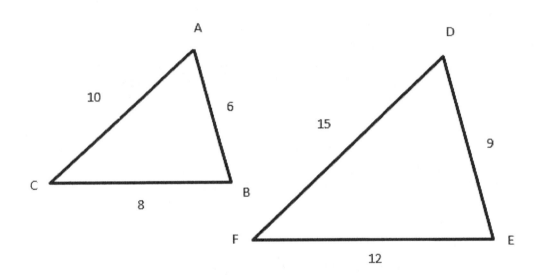

The triangles are similar because each ratio of corresponding sides is equal to 1.5.

Three methods can be used to show two triangles are congruent. First, if the three sides of the first triangle are equal to the sides of the second, the triangles are congruent. Equal sides means equal angles. Second, if two sides and their included angle of the first triangle can be shown to be equal to two sides and their included angle of the second triangle, the triangles are congruent. Third, if two angles and their included side of the first angle can be shown to be equal to two angles and their included side of the second angle, the triangles are congruent.

Relationships in geometric figures other than triangles can be proven using triangle congruence and similarity. If a similar or congruent triangle can be found within another type of geometric figure, their criteria can be used to prove a relationship about a given formula. For example, a rectangle can be broken up into two congruent triangles.

Three-Dimensional Figures

A **rectangular solid** is a six-sided figure with sides that are rectangles. All of the faces meet at right angles, and it looks like a box. Its three measurements are length l, width w, and height h.

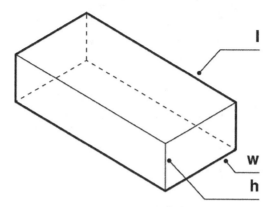

If all sides are equal in a rectangular solid, the solid is known as a **cube**. The cube has six congruent faces that meet at right angles, and each side length is the same and is labeled s.

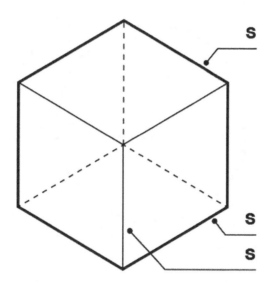

A **cylinder** is a three-dimensional geometric figure consisting of two parallel circles and two parallel lines connecting the ends. The circle has radius r, and the cylinder has height h.

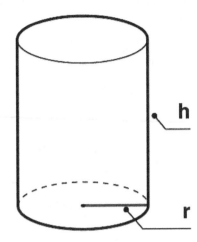

Finally, a **sphere** is a symmetrical three-dimensional shape, where every point on the surface is equal distance from its center. It has a radius r.

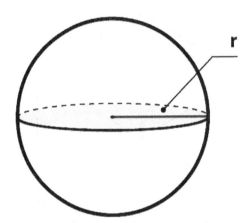

Coordinate Geometry

Coordinate geometry is the intersection of algebra and geometry. Within this system, the points in a geometric shape are defined using ordered pairs. In the two-dimensional coordinate system, an x- and y-axis form the xy-plane. The x-axis is a horizontal scale, and the y-axis is a vertical scale. The ordered pair where the axes cross is known as the **origin**. To the right of the origin, the x values are positive, and to the left of the origin, the x values are negative. The y values above the origin are positive, and y values below the origin are negative. The axes split the plane into four quadrants, and the first quadrant is where both x and y values are positive. To plot an ordered pair means to locate the point corresponding to the x and y coordinates. For example, plotting (4, 3) means moving to the right 4 units from zero in the x direction and then moving up 3 units in the y direction.

Here is a picture of the xy-plane, also known as the **rectangular** or **Cartesian coordinate system**:

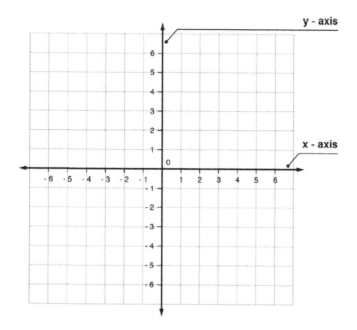

The coordinate system allows us to visualize relationships between equations and geometric figures. For instance, an equation in two variables, x and y, is represented as a straight line on the xy coordinate plane. A solution of an equation in two variables is an ordered pair that satisfies the equation. A graph of an equation can be found by plotting several ordered pairs that are solutions of the equation and then connecting those points with a straight line or smooth curve.

Here is the graph of $4x + y = 8$:

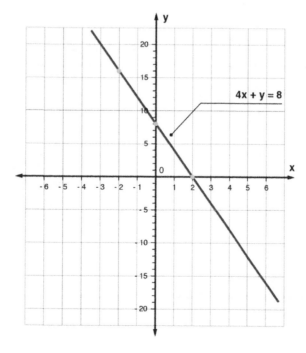

209

Three ordered pairs that are solutions to the equation were found and plotted. Those points are (–2, 16), (0, 8), and (2, 0). The points were connected using a straight line. Note that the point (0, 8) is where the line crosses the y-axis. This point is known as the y-intercept. The y-intercept can always be found by plugging $x = 0$ into the equation. Also, the point (2, 0) is where the line crosses the x-axis. This point is known as the x-intercept, and it can always be found for any equation of a line by plugging $y = 0$ into the equation. The equation above is written in standard form, $Ax + By = C$. Often an equation is written in slope-intercept form, $y = mx + b$, where m represents the slope of the line, and b represents the y-intercept. The above equation can be solved for y to obtain $y = -4x + 8$, which shows a slope of –4 and a y-intercept of 8, meaning the point (0, 8).

The slope of a line is the measure of steepness of a line, and it compares the vertical change of the line, the rise, to the horizontal change of the line, the run. The formula for slope of a line through two distinct points (x_1, y_1) and (x_2, y_2) is:

$$m = \frac{y_2 - y_1}{x_2 - x_1}$$

If the line increases from left to right, the slope is positive, and if the line decreases from left to right, as shown above, the slope is negative. If a line is horizontal, like the line representing the equation $y = 5$, the slope is zero. If a line is vertical, like the line representing the equation $x = 2$, the line has undefined slope.

In order to graph a function, it can be done the same way as equations. The $f(x)$ represents the dependent variable y in the equation, so replace $f(x)$ with y, and plot some points. For example, the same graph above would be found for the function:

$$f(x) = -4x + 8$$

Graphs other than straight lines also exist. For instance, here are the graphs of $f(x) = x^2$ and $f(x) = x^3$, the squaring and cubic functions.

f(x) = x²

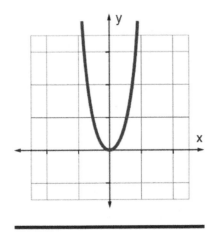

f(x) = x⁴

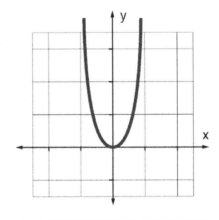

If the equals sign is changed to an inequality symbol such as $<, >, \leq,$ or $\geq$ in an equation, the result is an inequality. If it is changed to a linear equation in two variables, the result is a linear inequality in two variables. A solution of an inequality in two variables is an ordered pair that satisfies the inequality. For example, (1, 3) is a solution of the linear inequality $y \geq x + 1$ because when plugged in, it results in a true statement. The graph of an inequality in two variables consists of all ordered pairs that make the solution true.

A half-plane consists of the set of all points on one side of a line in the xy-plane, and the solution to a linear inequality is a half-plane. If the inequality consists of $>$ or $<$, the line is dashed, and no solutions actually exist on the line shown. If the inequality consists of $\geq$ or $\leq$, the line is solid, and solutions do exist on the line shown. In order to graph a linear inequality, graph the corresponding equation found by replacing the inequality symbol with an equals sign. Then pick a test point on either side of the line. If that point results in a true statement when plugged into the original inequality, shade in the side containing the test point. If it results in a false statement, shade in the opposite side. Here is the graph of the inequality $y < x + 1$.

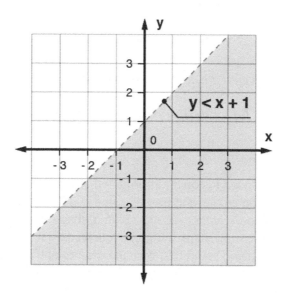

Conics

The intersection of a plane and a double right circular cone is called a **conic section**. There are four basic types of conic sections, a circle, a parabola, a hyperbola, and an ellipse. The equation of a circle is given by $(x - h)^2 + (y - k)^2 = r^2$, where the center of the circle is given by (h, k) and the radius of the circle is r. A parabola that opens up or down has a horizontal axis. The equation of a parabola with a horizontal axis is given by $(y - k)^2 = 4p(x - h)$, where $p \neq 0$ and the vertex is given by (h, k). A parabola that opens to the left or right has a vertical axis. The equation of the parabola with a vertical axis is given by $(x - h)^2 = 4p(y - k)$, where $p \neq 0$ and the vertex is given by (h, k). The equation of an ellipse with a horizontal major axis and center (h, k) is given by:

$$\frac{(x - h)^2}{a^2} + \frac{(y - k)^2}{b^2} = 1$$

The distance between center and either focus is c with $c^2 = a^2 - b^2$, when $a > b > 0$. The major axis has length $2a$ and the minor axis has length $2b$. For an ellipse with a vertical major axis and center (h, k), where $a > b > 0$, the a and b switch places so the equation is given by:

$$\frac{(x-h)^2}{b^2} + \frac{(y-k)^2}{a^2} = 1$$

The major axis still has length $2a$ and the minor axis still has length $2b$, and the distance between center and either focus is $c^2 = a^2 - b^2$, where $a > b > 0$.

A hyperbola has an equation similar to the ellipse except that there is a minus in place of the plus sign. A hyperbola with a vertical transverse axis has equation:

$$\frac{(x-h)^2}{a^2} - \frac{(y-k)^2}{b^2} = 1$$

A hyperbola with a horizontal transverse axis has equation:

$$\frac{(y-k)^2}{a^2} - \frac{(x-h)^2}{b^2} = 1$$

For each of these, the center is given by (h, k) and distance between the vertices $2a$.

Measurement and Data

Tables, Charts, and Graphs

Tables, charts, and graphs can be used to convey information about different variables. They are all used to organize, categorize, and compare data, and they all come in different shapes and sizes. Each type has its own way of showing information, whether it is in a column, shape, or picture. To answer a question relating to a table, chart, or graph, some steps should be followed. First, the problem should be read thoroughly to determine what is being asked to determine what quantity is unknown. Then, the title of the table, chart, or graph should be read. The title should clarify what actual data is being summarized in the table. Next, look at the key and both the horizontal and vertical axis labels, if they are given. These items will provide information about how the data is organized. Finally, look to see if there is any more labeling inside the table. Taking the time to get a good idea of what the table is summarizing will be helpful as it is used to interpret information.

Tables are a good way of showing a lot of information in a small space. The information in a table is organized in columns and rows. For example, a table may be used to show the number of votes each candidate received in an election. By interpreting the table, one may observe which candidate won the election and which candidates came in second and third. In using a bar chart to display monthly rainfall amounts in different countries, rainfall can be compared between countries at different times of the year. Graphs are also a useful way to show change in variables over time, as in a line graph, or percentages of a whole, as in a pie graph.

The table below relates the number of items to the total cost. The table shows that 1 item costs $5. By looking at the table further, 5 items cost $25, 10 items cost $50, and 50 items cost $250. This cost can be extended for any number of items. Since 1 item costs $5, then 2 items would cost $10. Though this information isn't in the table, the given price can be used to calculate unknown information.

Number of Items	1	5	10	50
Cost ($)	5	25	50	250

A **bar graph** is a graph that summarizes data using bars of different heights. It is useful when comparing two or more items or when seeing how a quantity changes over time. It has both a horizontal and vertical axis. Interpreting bar graphs includes recognizing what each bar represents and connecting that to the two variables. The bar graph below shows the scores for six people on three different games. The color of the bar shows which game each person played, and the height of the bar indicates their score for that game. William scored 25 on game 3, and Abigail scored 38 on game 3. By comparing the bars, it's obvious that Williams scored lower than Abigail.

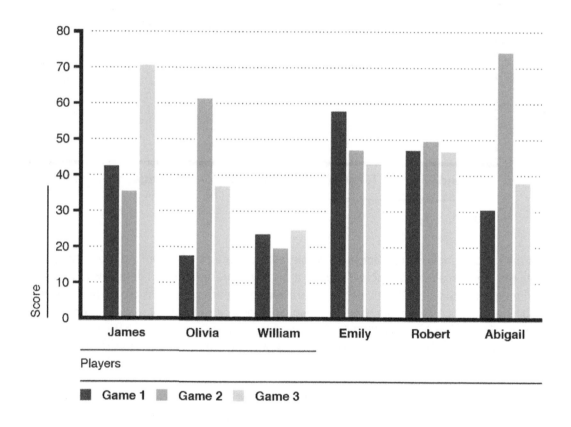

A **line graph** is a way to compare two variables. Each variable is plotted along an axis, and the graph contains both a horizontal and a vertical axis. On a line graph, the line indicates a continuous change. The change can be seen in how the line rises or falls, known as its slope, or rate of change. Often, in line graphs, the horizontal axis represents a variable of time. Audiences can quickly see if an amount has increased or decreased over time. The bottom of the graph, or the x-axis, shows the units for time, such as days, hours, months, etc. If there are multiple lines, a comparison can be made between what the two lines represent. For example, the following line graph shows the change in temperature over five days. The top line represents the high, and the bottom line represents the low for each day. Looking at the top line alone, the high decreases for a day, then increases on Wednesday. Then it decreases on Thursday and increases again on Friday. The low temperatures have a similar trend, shown in bottom line. The range in temperatures each day can also be calculated by finding the difference between the top line and bottom line on a particular day. On Wednesday, the range was 14 degrees, from 62 to 76° F.

Daily Temperatures

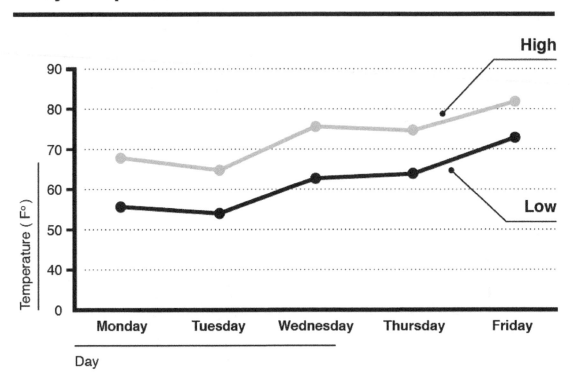

Pie charts are used to show percentages of a whole, as each category is given a piece of the pie, and together all the pieces make up a whole. They are a circular representation of data which are used to highlight numerical proportion. It is true that the arc length of each pie slice is proportional to the amount it individually represents. When a pie chart is shown, an audience can quickly make comparisons by comparing the sizes of the pieces of the pie. They can be useful for comparison between different categories. The following pie chart is a simple example of three different categories shown in comparison to each other.

Light gray represents cats, dark gray represents dogs, and the gray between those two represents other pets. As the pie is cut into three equal pieces, each value represents just more than 33 percent, or $\frac{1}{3}$ of the whole. Values 1 and 2 may be combined to represent $\frac{2}{3}$ of the whole.

In an example where the total pie represents 75,000 animals, then cats would be equal to $\frac{1}{3}$ of the total, or 25,000. Dogs would equal 25,000 and other pets also equal 25,000.

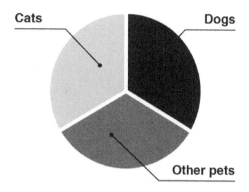

The fact that a circle is 360 degrees is used to create a pie chart. Because each piece of the pie is a percentage of a whole, that percentage is multiplied times 360 to get the number of degrees each piece represents. In the example above, each piece is $\frac{1}{3}$ of the whole, so each piece is equivalent to 120 degrees. Together, all three pieces add up to 360 degrees.

Stacked bar graphs, also used fairly frequently, are used when comparing multiple variables at one time. They combine some elements of both pie charts and bar graphs, using the organization of bar graphs and the proportionality aspect of pie charts. The following is an example of a stacked bar graph that represents the number of students in a band playing drums, flute, trombone, and clarinet. Each bar graph is broken up further into girls and boys:

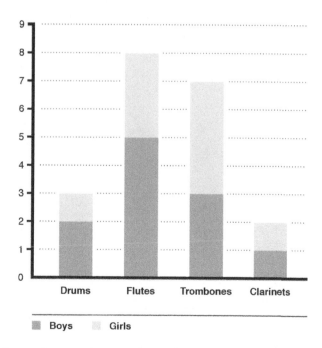

To determine how many boys play trombone, refer to the darker portion of the trombone bar, resulting in 3 students.

A **scatterplot** is another way to represent paired data. It uses Cartesian coordinates, like a line graph, meaning it has both a horizontal and vertical axis. Each data point is represented as a dot on the graph. The dots are never connected with a line. For example, the following is a scatterplot showing people's height versus age.

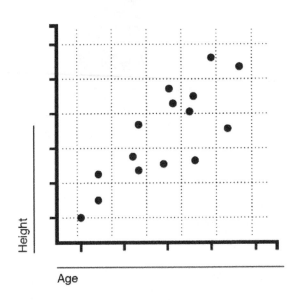

A scatterplot, also known as a **scattergram**, can be used to predict another value and to see if an association, known as **a correlation**, exists between a set of data. If the data resembles a straight line, the data **is associated**. The following is an example of a scatterplot in which the data does not seem to have an association:

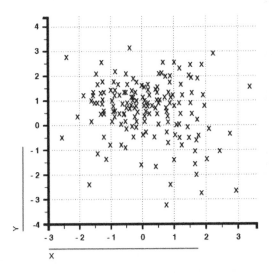

Sets of numbers and other similarly organized data can also be represented graphically. Venn diagrams are a common way to do so. A **Venn diagram** represents each set of data as a circle. The circles overlap, showing that each set of data is overlapping. A Venn diagram is also known as a **logic diagram** because it visualizes all possible logical combinations between two sets. Common elements of two sets are represented by the area of overlap.

The following is an example of a Venn diagram of two sets A and B:

Parts of the Venn Diagram

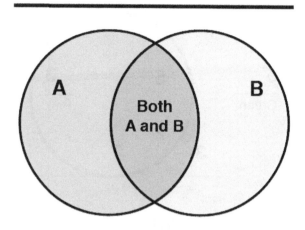

Another name for the area of overlap is the **intersection**. The intersection of A and B, $A \cap B$, contains all elements that are in both sets A and B. The **union** of A and B, $A \cup B$, contains all elements that are in either set A or set B. Finally, the **complement** of $A \cup B$ is equal to all elements that are not in either set A or set B. These elements are placed outside of the circles.

The following is an example of a Venn diagram in which 30 students were surveyed asking which type of siblings they had: brothers, sisters, or both. Ten students only had a brother, 7 students only had a sister, and 5 had both a brother and a sister. This number 5 is the intersection and is placed where the circles overlap. Two students did not have a brother or a sister. Two is therefore the complement and is placed outside of the circles.

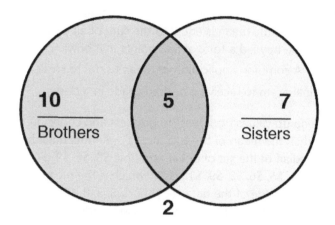

Venn diagrams can have more than two sets of data. The more circles, the more logical combinations are represented by the overlapping. The following is a Venn diagram that represents a different situation. Now, there were 30 students surveyed about the color of their socks. The innermost region represents

217

those students that have green, pink, and blue socks on (perhaps a striped pattern). Therefore, 2 students had all three colors on their socks. In this example, all students had at least one of the three colors on their socks, so no one exists in the complement.

30 students

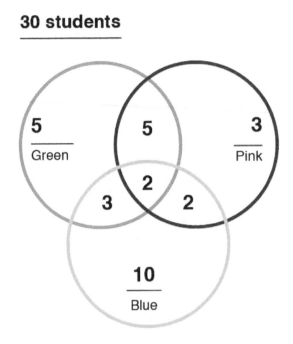

Venn diagrams are typically not drawn to scale, but if they are and their area is proportional to the amount of data it represents, it is known as an **area-proportional Venn diagram**.

Using the Relationship Between Two Variables to Investigate Key Features of the Graph

One way information can be interpreted from tables, charts, and graphs is through **statistics**. The three most common calculations for a set of data are the mean, median, and mode and are called **measures of central tendency**. Measures of central tendency are helpful in comparing two or more different sets of data. The **mean** refers to the average and is found by adding up all values and dividing the total by the number of values. In other words, the mean is equal to the sum of all values divided by the number of data entries. For example, if you bowled a total of 532 points in 4 bowling games, your mean score was $\frac{532}{4} = 133$ points per game. A common application of mean useful to students is calculating what he or she needs to receive on a final exam to receive a desired grade in a class.

The **median** is found by lining up values from least to greatest and choosing the middle value. If there's an even number of values, then the mean of the two middle amounts must be calculated to find the median. For example, the median of the set of dollar amounts $5, $6, $9, $12, and $13 is $9. The median of the set of dollar amounts $1, $5, $6, $8, $9, $10 is $7, which is the mean of $6 and $8. The **mode** is the value that occurs the most. The mode of the data set {1, 3, 1, 5, 5, 8, 10} actually refers to two numbers: 1 and 5. In this case, the data set is **bimodal** because it has two modes. A data set can have no mode if no amount is repeated. Another useful statistic is range. The **range** for a set of data refers to the difference between the highest and lowest value.

In some cases, some numbers in a list of data might have weights attached to them. In that case, a **weighted mean** can be calculated. A common application of a weighted mean is GPA. In a semester, each class is assigned a number of credit hours, its weight, and at the end of the semester each student receives a grade. To compute GPA, an A is a 4, a B is a 3, a C is a 2, a D is a 1, and an F is a 0. Consider a student that takes a 4-hour English class, a 3-hour math class, and a 4-hour history class and receives all B's. The weighted mean, GPA, is found by multiplying each grade times its weight, number of credit hours, and dividing by the total number of credit hours. Therefore, the student's GPA is:

$$\frac{3 \times 4 + 3 \times 3 + 3 \times 4}{11} = \frac{33}{1} = 3.0.$$

The following bar chart shows how many students attend a cycle class on each day of the week. To find the mean attendance for the week, each day's attendance can be added together, $10 + 7 + 6 + 9 + 8 + 14 + 4 = 58$, and the total divided by the number of days, $58 \div 7 = 8.3$. The mean attendance for the week was 8.3 people. The median attendance can be found by putting the attendance numbers in order from least to greatest: 4, 6, 7, 8, 9, 10, 14, and choosing the middle number: 8 people. There is no mode for this set of data because no numbers repeat. The range is 10, which is found by finding the difference between the lowest number, 4, and the highest number, 14.

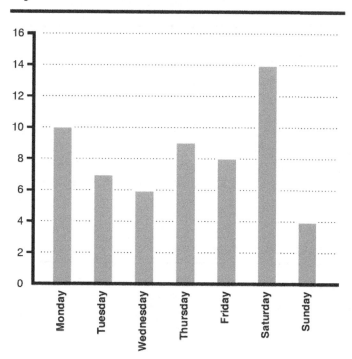

Cycle class attendance

A **histogram** is a bar graph used to group data into "bins" that cover a range on the horizontal, or x-axis. Histograms consist of rectangles whose height is equal to the frequency of a specific category. The horizontal axis represents the specific categories. Because they cover a range of data, these bins have no gaps between bars, unlike the bar graph above. In a histogram showing the heights of adult golden retrievers, the bottom axis would be groups of heights, and the y-axis would be the number of dogs in each range. Evaluating this histogram would show the height of most golden retrievers as falling within a certain range. It also provides information to find the average height and range for how tall golden retrievers may grow.

The following is a histogram that represents exam grades in a given class. The horizontal axis represents ranges of the number of points scored, and the vertical axis represents the number of students. For example, approximately 33 students scored in the 60 to 70 range.

Results of the exam

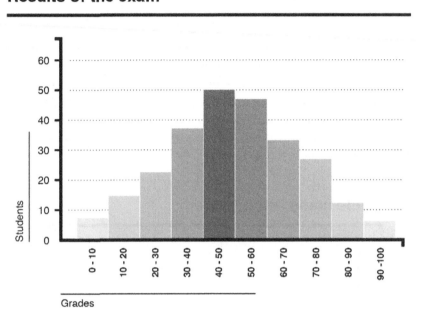

Measures of central tendency can be discussed using a histogram. If the points scored were shown with individual rectangles, the tallest rectangle would represent the mode. A bimodal set of data would have two peaks of equal height. Histograms can be classified as having data **skewed to the left**, **skewed to the right**, or **normally-distributed**, which is also known as **bell-shaped**. These three classifications can be seen in the following image:

Measures of central tendency images

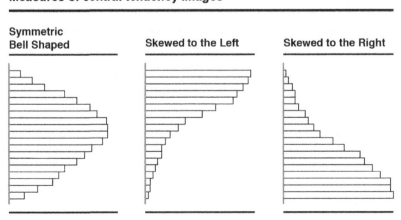

When the data follows the normal distribution, the mean, median, and mode are all very close. They all represent the most typical value in the data set. The mean is typically used as the best measure of central tendency in this case because it does include all data points. However, if the data is skewed, the mean becomes less meaningful. The median is the best measure of central tendency because it is not affected by any outliers, unlike the mean. When the data is skewed, the mean is dragged in the direction of the skew. Therefore, if the data is not normal, it is best to use the median as the measure of central tendency.

The measures of central tendency and the range may also be found by evaluating information on a line graph.

In the line graph from a previous example that showed the daily high and low temperatures, the average high temperature can be found by gathering data from each day on the triangle line. The days' highs are 82, 78, 75, 65, and 70. The average is found by adding them together to get 370, then dividing by 5 (because there are 5 temperatures). The average high for the five days is 74. If 74 degrees is found on the graph, then it falls in the middle of the values on the triangle line. The mean of the low temperature can be found in the same way.

Given a set of data, the **correlation coefficient**, r, measures the association between all the data points. If two values are **correlated**, there is an association between them. However, correlation does not necessarily mean **causation**, or that one value causes the other. There is a common mistake made that assumes correlation implies causation. Average daily temperature and number of sunbathers are both correlated and have causation. If the temperature increases, that change in weather causes more people to want to catch some rays. However, wearing plus-size clothing and having heart disease are two variables that are correlated but do not have causation. The larger someone is, the more likely he or she is to have heart disease. However, being overweight does not cause someone to have the disease.

The value of the correlation coefficient is between -1 and 1, where -1 represents a perfect negative linear relationship, zero represents no relationship between the two data sets, and 1 represents a perfect positive linear relationship. A **negative linear relationship** means that as x values increase, y values decrease. A **positive linear relationship** means that as x values increase, y values increase. The formula for computing the correlation coefficient is:

$$r = \frac{n \sum xy - (\sum x)(\sum y)}{\sqrt{n(\sum x^2) - (\sum x)^2}\sqrt{n(\sum y^2) - (y)^2}}$$

In this formula, n is the number of data points.

The closer r is to 1 or -1, the stronger the correlation. A correlation can be seen when plotting data. If the graph resembles a straight line, there is a correlation.

Standard Deviation, Quartiles, and Percentiles

A set of data can be described using its **standard deviation**, or spread. It measures how spread apart the data is within the set. The standard deviation actually quantifies the amount of variation with respect to the mean of the dataset. A lower standard deviation shows that the dataset does not differ much from the mean. A standard deviation equal to zero means that every value is the same in a dataset. Therefore, a larger standard deviation shows that the dataset, as a whole, varies largely from the mean.

The formula for sample standard deviation of a sample dataset is:

$$s = \sqrt{\frac{\sum(x - \bar{x})^2}{n - 1}}$$

where x is each value in the dataset, $\bar{x}$ is the mean, and n is the total number of data points in the set.

A dataset can be broken up into four equal parts. The three **quartiles** Q_1, Q_2, and Q_3 split up the data into four equal parts. Q_1 is the first quartile, and one-quarter of the data falls on or below it. Q_2 is the second quartile, also the median, and one-half of the data falls on or below it. Q_3 is the third quartile, and three-quarters of the data falls on or below it. The **interquartile range** *(IQR)* of a dataset gives the range of the middle 50 percent of the data, and its formula is:

$$IQR = Q_3 - Q_1$$

Similar to quartiles, **deciles** divide a dataset into ten equal parts, and **percentiles** divide a dataset into one hundred equal parts. For example, the 90th percentile refers to splitting up a dataset into the bottom 90 percent of the data and the top 10 percent of the data.

Constructing Graphs That Correctly Represent Given Data

As mentioned, data is often displayed with a line graph, bar graph, or pie chart.

The line graph below shows the number of push-ups that a student did over one week:

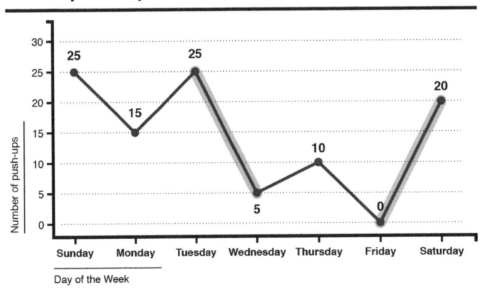

Notice that the horizontal axis displays the day of the week and the vertical axis displays the number of push-ups. A point is placed above each day of the week to show how many push-ups were done each day. For example, on Sunday the student did 25 push-ups. The line that connects the points shows how much the number of push-ups fluctuated throughout the week.

The bar graph below compares the number of people who own various types of pets:

What kind of pet do you own?

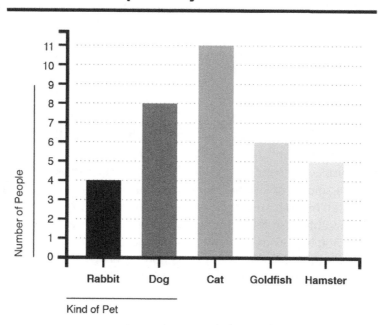

On the horizontal axis, the kind of pet is displayed. On the vertical axis, the number of people is displayed. Bars are drawn to show the number of people who own each type of pet. With the bar graph, it can quickly be determined that the fewest number of people own a rabbit and the greatest number of people own a cat.

The pie graph below displays students in a class who scored A, B, C, or D. Each slice of the pie is drawn to show the portion of the whole class that is represented by each letter grade. For example, the smallest portion represents students who scored a D. This means that the fewest number of students scored a D.

Student Grades

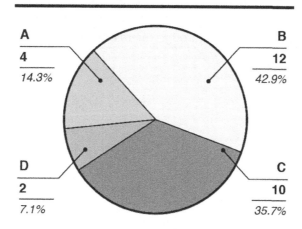

In summary:

- A line graph is used to display data that changes continuously over time.

- A bar graph is used to compare data from different categories or groups and is helpful for recognizing relationships.

- A pie chart is used when the data represents parts of a whole.

Important Features of Graphs

A **graph** is a pictorial representation of the relationship between two variables. To read and interpret a graph, it is necessary to identify important features of the graph. First, read the title to determine what data sets are being related in the graph. Next, read the axis labels and understand the scale that is used. The horizontal axis often displays categories, like years, month, or types of pets. The vertical axis often displays numerical data like amount of income, number of items sold, or number of pets owned. Check to see what increments are used on each axis. The changes on the axis may represent fives, tens, hundreds, or any increment. Be sure to note what the increment is because it will affect the interpretation of the graph. Now, locate on the graph an element of interest and move across to find the element to which it relates. For example, notice an element displayed on the horizontal axis, find that element on the graph, and then follow it across to the corresponding point on the vertical axis. Using the appropriate scale, interpret the relationship.

Explaining the Relationship between Two Variables

Independent and dependent are two types of variables that describe how they relate to each other. The **independent variable** is the variable controlled by the experimenter. It stands alone and isn't changed by other parts of the experiment. This variable is normally represented by x and is found on the horizontal, or x-axis, of a graph. The **dependent variable** changes in response to the independent variable. It reacts to, or depends on, the independent variable. This variable is normally represented by y and is found on the vertical, or y-axis of the graph.

The relationship between two variables, x and y, can be seen on a scatterplot.

The following scatterplot shows the relationship between weight and height. The graph shows the weight as x and the height as y. The first dot on the left represents a person who is 45 kg and approximately 150 cm tall. The other dots correspond in the same way. As the dots move to the right and weight increases, height also increases. A line could be drawn through the middle of the dots to move from bottom left to top right. This line would indicate a **positive correlation** between the variables. If the variables had a **negative correlation**, then the dots would move from the top left to the bottom right.

Height and Weight

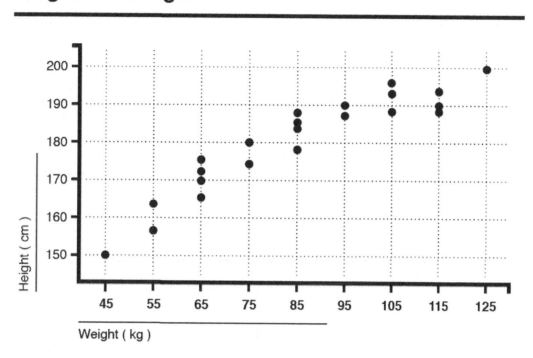

A **scatterplot** is useful in determining the relationship between two variables, but it's not required. Consider an example where a student scores a different grade on his math test for each week of the month. The independent variable would be the weeks of the month. The dependent variable would be the grades, because they change depending on the week. If the grades trended up as the weeks passed, then the relationship between grades and time would be positive. If the grades decreased as the time passed, then the relationship would be negative. (As the number of weeks went up, the grades went down.)

The relationship between two variables can further be described as strong or weak. The relationship between age and height shows a strong positive correlation because children grow taller as they grow up. In adulthood, the relationship between age and height becomes weak, and the dots will spread out. People stop growing in adulthood, and their final heights vary depending on factors like genetics and health. The closer the dots on the graph, the stronger the relationship. As they spread apart, the relationship becomes weaker. If they are too spread out to determine a correlation up or down, then the variables are said to have no correlation.

Variables are values that change, so determining the relationship between them requires an evaluation of who changes them. If the variable changes because of a result in the experiment, then it's dependent. If the variable changes before the experiment, or is changed by the person controlling the experiment, then it's the independent variable. As they interact, one is manipulated by the other. The manipulator is the independent, and the manipulated is the dependent. Once the independent and dependent variable are determined, they can be evaluated to have a positive, negative, or no correlation.

Comparing Linear Growth with Exponential Growth

Linear growth involves a quantity, the dependent variable, increasing or decreasing at a constant rate as another quantity, the independent variable, increases as well. The graph of linear growth is a straight line. Linear growth is represented as the following equation: $y = mx + b$, where m is the slope of the line, also known as the rate of change, and b is the y-intercept. If the y-intercept is zero, then the linear growth is actually known as direct variation. If the slope is positive, the dependent variable increases as the independent variable increases, and if the slope is negative, the dependent variable decreases as the independent variable increases.

Exponential growth involves a quantity, the dependent variable, changing by a common ratio every unit increase or equal interval. The equation of exponential growth is $y = a^x$ for $a > 0$, $a \neq 1$. The value a is known as the **base**. Consider the exponential equation $y = 2^x$. When x equals 1, y equals 2, and when x equals 2, y equals 4. For every unit increase in x, the value of the output variable doubles. Here is the graph of $y = 2^x$. Notice that as the dependent variable, y, gets very large, x increases slightly. This characteristic of this graph is why sometimes a quantity is said to be blowing up exponentially.

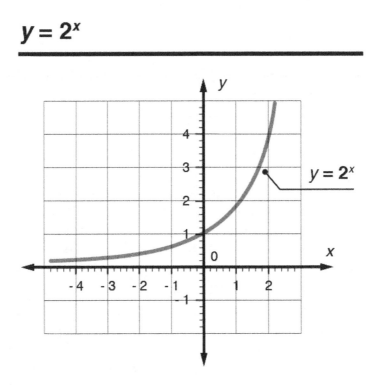

Statistics and Probability

Statistical Questions

Statistics is the branch of mathematics that deals with the collection, organization, and analysis of data. A statistical question is one that can be answered by collecting and analyzing data. When collecting data, expect variability. For example, "How many pets does Yanni own?" is not a statistical question because it can be answered in one way. "How many pets do the people in a certain neighborhood own?" is a statistical question because, to determine this answer, one would need to collect data from each person in the neighborhood, and it is reasonable to expect the answers to vary.

Identify these as statistical or not statistical:

1. How old are you?

2. What is the average age of the people in your class?

3. How tall are the students in Mrs. Jones' sixth grade class?

4. Do you like Brussels sprouts?

Questions 2 and 3 are statistical questions.

How Changes in Data Affect Measures of Center or Range

An **outlier** is a data point that lies an unusual distance from other points in the data set. Removing an outlier from a data set will change the measures of center. Removing a large outlier from a data set will decrease both the mean and the median. Removing a small outlier from a data set will increase both the mean and the median. For example, given the data set {3, 6, 8, 12, 13, 14, 60}, the data point 60 is an outlier because it is unusually far from the other points. In this data set, the mean is 16.6. Notice that this mean number is even larger than all other data points in the set except for 60. Removing the outlier, the mean changes to 9.3 and the median becomes 10. Removing an outlier will also decrease the range. In the data set above, the range is 57 when the outlier is included, but decreases to 11 when the outlier is removed.

Adding an outlier to a data set will affect the centers of measure as well. When a larger outlier is added to a data set, the mean and median increase. When a small outlier is added to a data set, the mean and median decrease. Adding an outlier to a data set will increase the range.

This does not seem to provide an appropriate measure of center when considering this data set. What will happen if that outlier is removed? Removing the extremely large data point, 60, is going to reduce the mean to 9.3. The mean decreased dramatically because 60 was much larger than any of the other data points. What would happen with an extremely low value in a data set like this one, {12, 87, 90, 95, 98, 100}? The mean of the given set is 80. When the outlier, 12, is removed, the mean should increase and should fit more closely to the other data points. Removing 12 and recalculating the mean shows that this is correct. The mean after removing 12 is 94. So, removing a large outlier will decrease the mean, while removing a small outlier will increase the mean.

Data Collection Methods

Data collection can be done through surveys, experiments, observations, and interviews. A **census** is a type of survey that is done with a whole population. Because it can be difficult to collect data for an entire population, sometimes a **sample survey** is used. In this case, one would survey only a fraction of the population and make inferences about the data. Sample surveys are not as accurate as a census, but this is an easier and less expensive method of collecting data. An **experiment** is used when a researcher wants to explain how one variable causes changes in another variable. For example, if a researcher wanted to know if a particular drug affects weight loss, he would choose a treatment group that would take the drug, and another group, the control group, that would not take the drug. Special care must be taken when choosing these groups to ensure that bias is not a factor. **Bias** occurs when an outside factor influences the outcome of the research. In observational studies, the researcher does not try to influence either variable but simply observes the behavior of the subjects. Interviews are sometimes used to collect data as well. The researcher will ask questions that focus on her area of interest in order to gain insight from the participants. When gathering data through observation or interviews, it is important that the researcher is well trained so that he does not influence the results and so that the study is reliable. A study is reliable if it can be repeated under the same conditions and the same results are obtained each time.

Making Inferences About Population Parameters Based on Sample Data

In statistics, a **population** contains all subjects being studied. For example, a population could be every student at a university or all males in the United States. A **sample** consists of a group of subjects from an entire population. A sample would be 100 students at a university or 100,000 males in the United States. **Inferential statistics** is the process of using a sample to generalize information concerning populations. **Hypothesis testing** is the actual process used when evaluating claims made about a population based on a sample.

A **statistic** is a measure obtained from a sample, and a **parameter** is a measure obtained from a population. For example, the mean SAT score of the 100 students at a university would be a statistic, and the mean SAT score of all university students would be a parameter.

The beginning stages of hypothesis testing starts with formulating a **hypothesis**, a statement made concerning a population parameter. The hypothesis may be true, or it may not be true. The test will answer that question. In each setting, there are two different types of hypotheses: the **null hypothesis**, written as H_0, and the **alternative hypothesis**, written as H_1. The null hypothesis represents verbally when there is not a difference between two parameters, and the alternative hypothesis represents verbally when there is a difference between two parameters.

Consider the following experiment: A researcher wants to see if a new brand of allergy medication has any effect on drowsiness of the patients who take the medication. He wants to know if the average hours spent sleeping per day increases. The mean for the population under study is 8 hours, so $\mu = 8$. In other words, the population parameter is μ, the mean. The null hypothesis is $\mu = 8$ and the alternative hypothesis is $\mu > 8$. When using a smaller sample of a population, the null hypothesis represents the situation when the mean remains unaffected and the alternative hypothesis represents the situation when the mean increases. The chosen statistical test will apply the data from the sample to actually decide whether the null hypothesis should or should not be rejected.

Dependent Versus Independent Variables

Independent variables are independent, meaning they are not changed by other variables within the context of the problem. **Dependent variables** are dependent, meaning they may change depending on how other variables change in the problem. For example, in the formula for the perimeter of a fence, the length and width are the independent variables and the perimeter is the dependent variable. The formula is shown below:

$$P = 2l + 2w$$

As the width or the length changes, the perimeter may also change. The first variables to change are the length and width, which then result in a change in perimeter. The change does not come first with the perimeter and then with length and width. When comparing these two types of variables, it is helpful to ask which variable causes the change and which variable is affected by the change.

Another formula to represent this relationship is the formula for circumference shown below:

$$C = \pi \times d$$

The C represents circumference and the d represents diameter. The pi symbol is approximated by the fraction $\frac{22}{7}$, or 3.14. In this formula, the diameter of the circle is the independent variable. It is the portion of the circle that changes, which changes the circumference as a result. The circumference is the variable that is being changed by the diameter, so it is called the dependent variable. It depends on the value of the diameter.

Another place to recognize independent and dependent variables can be in experiments. A common experiment is one where the growth of a plant is tested based on the amount of sunlight it receives. Each plant in the experiment is given a different amount of sunlight, but the same amount of other nutrients like light and water. The growth of the plants is measured over a given time period and the results show how much sunlight is best for plants. In this experiment, the independent variable is the amount of sunlight that each plant receives. The dependent variable is the growth of each plant. The growth depends on the amount of sunlight, which gives reason for the distinction between independent and dependent variables.

Interpreting Probabilities Relative to Likelihood of Occurrence

Probability describes how likely it is that an event will occur. Probabilities are always a number from zero to 1. If an event has a high likelihood of occurrence, it will have a probability close to 1. If there is only a small chance that an event will occur, the likelihood is close to zero. A fair six-sided die has one of the numbers 1, 2, 3, 4, 5, and 6 on each side. When this die is rolled there is a one in six chance that it will land on 2. This is because there are six possibilities and only one side has a 2 on it. The probability then is $\frac{1}{6}$ or 0.167. The probability of rolling an even number from this die is three in six, which is $\frac{1}{2}$ or 0.5. This is because there are three sides on the die with even numbers (2, 4, 6), and there are six possible sides. The probability of rolling a number less than 10 is one because every side of the die has a number less than 6, so this is certain to occur. On the other hand, the probability of rolling a number larger than 20 is zero. There are no numbers greater than 20 on the die, so it is certain that this will not occur, thus the probability is zero.

If a teacher says that the probability of anyone passing her final exam is 0.2, is it highly likely that anyone will pass? No, the probability of anyone passing her exam is low because 0.2 is closer to zero than to 1. If another teacher is proud that the probability of students passing his class is 0.95, how likely is it that a student will pass? It is highly likely that a student will pass because the probability, 0.95, is very close to 1.

Elementary Probability

A **probability experiment** is an action that causes specific results, such as counts or measurements. The result of such an experiment is known as an **outcome**, and the set of all potential outcomes is known as the **sample space**. An **event** consists of one or more of those outcomes. For example, consider the probability experiment of tossing a coin and rolling a six-sided die. The coin has two possible outcomes—a heads or a tails—and the die has six possible outcomes—rolling each number 1–6. Therefore, the sample space has twelve possible outcomes: a heads or a tails paired with each roll of the die.

A **simple event** is an event that consists of a single outcome. For instance, selecting a queen of hearts from a standard fifty-two-card deck is a simple event; however, selecting a queen is not a simple event because there are four possibilities.

Classical probability, also known as theoretical probability, is when each outcome in a sample space has the same chance to occur. The probability for an event is equal to the number of outcomes in that event divided by the total number of outcomes in the sample space. For example, consider rolling a six-sided die. The probability of rolling a 2 is $\frac{1}{6}$, and the probability of rolling an even number is $\frac{3}{6}$, or $\frac{1}{2}$, because there are three even numbers on the die. This type of probability is based on what should happen in theory but not what actually happens in real life.

Empirical probability is based on actual experiments or observations. For instance, if a die is rolled eight times, and a 1 is rolled two times, the empirical probability of rolling a 1 is $\frac{2}{8} = \frac{1}{4}$, which is higher than the theoretical probability. The Law of Large Numbers states that as an experiment is completed repeatedly, the empirical probability of an event should get closer to the theoretical probability of an event.

Probabilities range from zero to 1. The closer the probability of an event occurring is to 0, the less likely it will occur. The closer it is to 1, the more likely it is to occur.

The **addition rule** is necessary to find the probability of event A or event B occurring or both occurring at the same time. If events A and B are **mutually exclusive** or **disjoint**, which means they cannot occur at the same time, $P(A \text{ or } B) = P(A) + P(B)$. If events A and B are not **mutually** exclusive, $P(A \text{ or } B) = P(A) + P(B) - P(A \text{ and } B)$ where $P(A \text{ and } B)$ represents the probability of event A and B both occurring at the same time. An example of two events that are mutually exclusive are rolling a 6 on a die and rolling an odd number on a die. The probability of rolling a 6 or rolling an odd number is:

$$\frac{1}{6} + \frac{3}{6} = \frac{4}{6} = \frac{2}{3}$$

Rolling a 6 and rolling an even number are not mutually exclusive because there is some overlap. The probability of rolling a 6 or rolling an even number is:

$$\frac{1}{6} + \frac{3}{6} - \frac{1}{6} = \frac{3}{6} = \frac{1}{2}$$

Conditional Probability
The **multiplication rule** is necessary when finding the probability that an event A occurs in a first trial and event B occurs in a second trial, which is written as $P(A \text{ and } B)$. This rule differs if the events are independent or dependent. Two events A and B are independent if the occurrence of one event does not affect the probability that the other will occur. If A and B are not independent, they are dependent, and the outcome of the first event somehow affects the outcome of the second. If events A and B are independent, $P(A \text{ and } B) = P(A)P(B)$, and if events A and B are dependent, $P(A \text{ and } B) = P(A)P(B|A)$, where $P(B|A)$ represents the probability event B occurs given that event A has already occurred.

$P(B|A)$ represents **conditional probability**, or the probability of event B occurring given that event A has already occurred. $P(B|A)$ can be found by dividing the probability of events A and B both occurring by the probability of event A occurring using the formula $P(B|A) = \frac{P(A \text{ and } B)}{P(A)}$ and represents the total number of outcomes remaining for B to occur after A occurs. This formula is derived from the multiplication rule with dependent events by dividing both sides by $P(A)$. Note that $P(B|A)$ and $P(A|B)$ are not the same. The first quantity shows that event B has occurred after event A, and the second quantity shows that event A has occurred after event B. To incorrectly interchange these ideas is known as **confusion of the inverse**.

Consider the case of drawing two cards from a deck of fifty-two cards. The probability of pulling two queens would vary based on whether the initial card was placed back in the deck for the second pull. If the card is placed back in, the probability of pulling two queens is:

$$\frac{4}{52} \times \frac{4}{52} = 0.00592$$

If the card is not placed back in, the probability of pulling two queens is:

$$\frac{4}{52} \times \frac{3}{51} = 0.00452$$

When the card is not placed back in, both the numerator and denominator of the second probability decrease by 1. This is due to the fact that, theoretically, there is one less queen in the deck, and there is one less total card in the deck as well.

Conditional probability is used frequently when probabilities are calculated from tables. Two-way frequency tables display data with two variables and highlight the relationships between those two variables. They are often used to summarize survey results and are also known as **contingency tables**. Each cell shows a count pertaining to that individual variable pairing, known as a **joint frequency**, and the totals of each row and column also are in the tables. Consider the following two-way frequency table:

	70 or older	**69 or younger**	**Totals**
Women	20	40	60
Men	5	35	40
Total	25	75	100

This table shows the breakdown of ages and sexes of 100 people in a particular village. Consider a randomly selected villager. The probability of selecting a male 69 years old or younger is $\frac{35}{100}$ because there are 35 males under the age of 70 and 100 total villagers.

Probability Distributions
A **discrete random variable** is a set of values that is either finite or countably infinite. If there are infinitely many values, being **countable** means that each individual value can be paired with a natural number. For example, the number of coin tosses before getting heads could potentially be infinite, but the total number of tosses is countable. Each toss refers to a number, like the first toss, second toss, etc. A **continuous random variable** has infinitely many values that are not countable. The individual items cannot be enumerated; an example of such a set is any type of measurement.

There are infinitely many heights of human beings due to decimals that exist within each inch, centimeter, millimeter, etc. Each type of variable has its own **probability distribution**, which calculates the probability for each potential value of the random variable. Probability distributions exist in tables, formulas, or graphs. The expected value of a random variable represents what the mean value should be in either a large sample size or after many trials. According to the Law of Large Numbers, after many trials, the actual mean and that of the probability distribution should be very close to the expected value. The **expected value** is a weighted average that is calculated as $E(X) = \sum x_i p_i$, where x_i represents the value of each outcome, and p_i represents the probability of each outcome.

The expected value if all of the probabilities are equal is:

$$E(X) = \frac{x_1 + x_2 + \cdots + x_n}{n}$$

Expected value is often called the **mean of the random variable** and is known as a **measure of central tendency** like mean and mode.

A **binomial probability distribution** is a probability distribution that adheres to some important criteria. The distribution must consist of a fixed number of trials where all trials are independent, each trial has an outcome classified as either success or failure, and the probability of a success is the same in each trial. Within any binomial experiment, x is the number of resulting successes, n is the number of trials, P is the probability of success within each trial, and $Q = 1 - P$ is the probability of failure within each trial. The probability of obtaining x successes within n trials is:

$\binom{n}{x} P^x (1 - P)^{n-x}$, where $\binom{n}{x} = \frac{n!}{x!(n-x)!}$ is called the **binomial coefficient**. A binomial probability distribution could be used to find the probability of obtaining exactly two heads on five tosses of a coin. In the formula, $x = 2$, $n = 5$, $P = 0.5$, and $Q = 0.5$.

A **uniform probability distribution** exists when there is constant probability. Each random variable has equal probability, and its graph is a rectangle because the height, representing the probability, is constant.

Finally, a **normal probability distribution** has a graph that is symmetric and bell-shaped; an example using body weight is shown here:

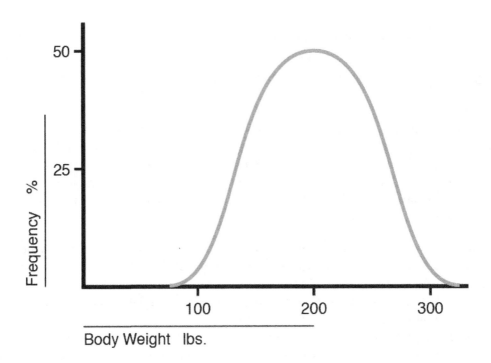

Population percentages can be estimated using normal distributions. For example, the probability that a data point will be less than the mean is 50 percent. The Empirical Rule states that 68 percent of the data falls within 1 standard deviation of the mean, 95 percent falls within 2 standard deviations of the mean, and 99.7 percent falls within 3 standard deviations of the mean. A **standard normal distribution** is a normal distribution with a mean equal to zero and standard deviation equal to 1. The area under the entire curve of a standard normal distribution is equal to 1.

Practice Questions

1. What type of units are used to describe surface area?
    a. Square
    b. Cubic
    c. Linear
    d. Quartic

2. How is a transposition of a matrix performed?
    a. Multiply each number by negative 1
    b. Switch the rows and columns
    c. Reverse the order of each row
    d. Find the inverse of each number

3. Which of the following is equivalent to $16^{\frac{1}{4}} \times 16^{\frac{1}{2}}$?
    a. 8
    b. 16
    c. 4
    d. 4,096

4. Which of the following values could NOT represent a probability?
    a. 0.0123

    b. 0.99999

    c. $\frac{1}{10}$

    d. $\frac{3}{2}$

5. Which of the following is the correct factorization of the polynomial $625x^8 - 25y^4$?
    a. $25(5x^4 + y^2)(5x^2 - y)(x^2 + y)$

    b. $25(5x^4 + y^2)(5x^4 - y^2)$

    c. $(25x^4 + 5y^2)(25x^4 - 5y^2)$

    d. $(25x^4 - y^2)(5x^4 - 5y^2)$

6. Which of the following is the result of the expression $\frac{14x}{y^2} - \frac{10y}{x^2}$ evaluated for $x = -2$ and $y = -3$?
    a. $\frac{79}{18}$

    b. $\frac{191}{18}$

    c. $-\frac{191}{18}$

    d. $-\frac{79}{18}$

7. If $4^x = 1024$, what is the value of $9x^4$?

    a. 5

    b. 5,625

    c. 45

    d. 625

8. Which of the following represents the equation $z = \frac{xy-4}{x-y}$ when solved for y?

    a. $y = \frac{xz+4}{z+x}$

    b. $y = \frac{z+x}{xz+4}$

    c. $y = \frac{x-y}{xy-4}$

    d. $y = (xz + 4)(z + x)$

9. If x is directly proportional to y and if $x = 8$ when $y = \frac{1}{12}$, then which of the following describes an equation that relates x and y?

    a. $x = 96y$

    b. $y = 96x$

    c. $x = 12y$

    d. $y = 1.5x$

10. If the ratio of x to y is 1:8, what is the product of x and y when $y = 48$?

    a. 96

    b. 183

    c. 48

    d. 288

11. The percent increase from 8 to 18 is equivalent to the percent increase from 234 to what number?

    a. 468

    b. 526

    c. 526.5

    d. 125

12. On the first four tests this semester, a student received the following scores out of 100: 74, 76, 82, and 84. The student must earn at least what score on the fifth test to receive a B in the class? Assume that the final test is also out of 100 points and that to receive a B in the class, he must have at least an 80% average.

    a. 80

    b. 84

    c. 76

    d. 78

13. In a class of 3 girls and 5 boys, the average score for boys was an 82 and the average for girls was 84. What was the class average?

    a. 82

    b. 82.75

    c. 83

    d. 83.75

14. How many 5-letter orderings of the word FRANCHISE can be made if no letters are repeated?

    a. 3,024

    b. 504

    c. 72

    d. 15,120

15. A set of cards contains n numbers, one of which is an odd number. If one card is randomly selected from the set, what is the probability that the card is even?

    a. $\dfrac{1}{n}$

    b. $\dfrac{1}{n-1}$

    c. $\dfrac{n-2}{n-1}$

    d. $\dfrac{n-1}{n}$

16. At a school raffle, 1000 tickets are sold for $1 each with the possibility of winning a $200 or a $50 prize. What is the expected value of the gain?

    a. $0.75

    b. $1

    c. -$0.75

    d. $1.25

17. Which of the following pairs of angles could NOT be the smaller and larger interior angles of a parallelogram?

    a. 120°, 60°

    b. 125°, 55°

    c. 110°, 60°

    d. 20°, 160°

18. What is the solution to the equation $10 - 5x + 2 = 7x + 12 - 12x$?

    a. $x = 1$

    b. $x = 0$

    c. All real numbers

    d. There is no solution

19. Which of the following is the result when solving the equation $4(x + 5) + 6 = 2(2x + 3)$?

    a. $x = 26$

    b. $x = 6$

    c. All real numbers

    d. There is no solution

20. Two consecutive integers exist such that the sum of three times the first and two less than the second is equal to 411. What are those integers?

    a. 103 and 104

    b. 104 and 105

    c. 102 and 103

    d. 100 and 101

21. In a neighborhood, 15 out of 80 of the households have children under the age of 18. What percentage of the households have children?
    a. 0.1875%
    b. 18.75%
    c. 1.875%
    d. 15%

22. If a car is purchased for $15,395 with a 7.25% sales tax, what is the total price?
    a. $15,395.07
    b. $16,511.14
    c. $16,411.13
    d. $15,402

23. From the chart below, which two are preferred by more men than women?

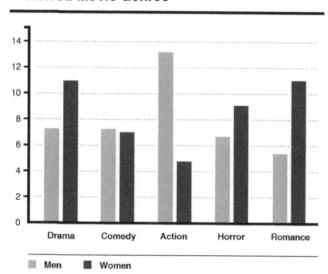

Preferred Movie Genres

    a. Comedy and Action
    b. Drama and Comedy
    c. Action and Horror
    d. Action and Romance

24. Which type of graph best represents a continuous change over a period of time?
    a. Bar graph
    b. Line graph
    c. Pie graph
    d. Histogram

25. Using the graph below, what is the mean number of visitors for the first 4 hours?

Museum Visitors

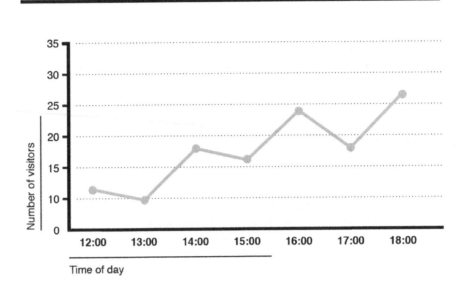

Time of day

a. 12
b. 13
c. 14
d. 15

26. What is the mode for the grades shown in the chart below?

Science Grades	
Jerry	65
Bill	95
Anna	80
Beth	95
Sara	85
Ben	72
Jordan	98

a. 65
b. 33
c. 95
d. 90

27. What type of relationship is there between age and attention span as represented in the graph below?

Attention Span

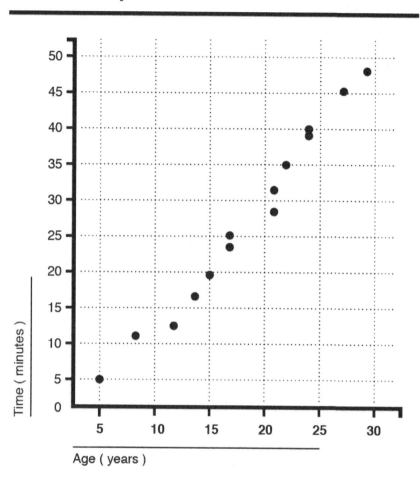

a. No correlation
b. Positive correlation
c. Negative correlation
d. Weak correlation

28. How many kiloliters are in 6 liters?
a. 6,000
b. 600
c. 0.006
d. 0.0006

29. Which of the following relations is a function?
   a. {(1, 4), (1, 3), (2, 4), (5, 6)}
   b. {(-1, -1), (-2, -2), (-3, -3), (-4, -4)}
   c. {(0, 0), (1, 0), (2, 0), (1, 1)}
   d. {(1, 0), (1, 2), (1, 3), (1, 4)}

30. Find the indicated function value: $f(5)$ for $f(x) = x^2 - 2x + 1$.
   a. 16
   b. 1
   c. 5
   d. Does not exist

31. What is the domain of $f(x) = 4x^2 + 2x - 1$?
   a. $(0, \infty)$
   b. $(-\infty, 0)$
   c. $(-\infty, \infty)$
   d. $(-1, 4)$

32. What is the range of the polynomial function $f(x) = 2x^2 + 5$?
   a. $(-\infty, \infty)$
   b. $(2, \infty)$
   c. $(0, \infty)$
   d. $[5, \infty)$

33. For which two values of x are the following functions equal?
$$f(x) = 4x + 4$$
$$g(x) = x^2 + 3x + 2$$
   a. 1, 0
   b. -2, -1
   c. -1, 2
   d. 1, 2

34. The population of coyotes in the local national forest has been declining since 2000. The population can be modeled by the function $y = -(x - 2)^2 + 1600$, where y represents number of coyotes and x represents the number of years past 2000. When will there be no more coyotes?
   a. 2020
   b. 2040
   c. 2012
   d. 2042

35. A ball is thrown up from a building that is 800 feet high. Its position s in feet above the ground is given by the function $s = -32t^2 + 90t + 800$, where t is the number of seconds since the ball was thrown. How long will it take for the ball to come back to its starting point? Round your answer to the nearest tenth of a second.
   a. 0 seconds
   b. 2.8 seconds
   c. 3 seconds
   d. 8 seconds

36. What is the domain of the following rational function?
$$f(x) = \frac{x^3 + 2x + 1}{2 - x}$$

    a. $(-\infty, -2) \cup (-2, \infty)$
    b. $(-\infty, 2) \cup (2, \infty)$
    c. $(2, \infty)$
    d. $(-2, \infty)$

37. What is the missing length x?

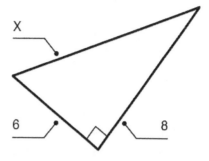

    a. 6
    b. 14
    c. 10
    d. 100

38. A study of adult drivers finds that it is likely that an adult driver wears his seatbelt. Which of the following could be the probability that an adult driver wears his seat belt?
    a. 0.90
    b. 0.05
    c. 0.25
    d. 0

39. What is the solution to the following linear inequality?
$$7 - \frac{4}{5}x < \frac{3}{5}$$

    a. $(-\infty, 8)$
    b. $(8, \infty)$
    c. $[8, \infty)$
    d. $(-\infty, 8]$

40. What is the solution to the following system of linear equations?
$$2x + y = 14$$
$$4x + 2y = -28$$

    a. (0, 0)
    b. (14, -28)
    c. All real numbers
    d. There is no solution

41. Which of the following is perpendicular to the line $4x + 7y = 23$?

a. $y = -\frac{4}{7}x + 23$

b. $y = \frac{7}{4}x - 12$

c. $4x + 7y = 14$

d. $y = -\frac{7}{4}x + 11$

42. What is the solution to the following system of equations?
$$2x - y = 6$$
$$y = 8x$$

a. (1, 8)
b. (-1, 8)
c. (-1, -8)
d. There is no solution.

43. The mass of the moon is about 7.348×10^{22} kilograms and the mass of Earth is 5.972×10^{24} kilograms. How many times GREATER is Earth's mass than the moon's mass?

a. 8.127×10^1
b. 8.127
c. 812.7
d. 8.127×10^{-1}

44. The percentage of smokers above the age of 18 in 2000 was 23.2 percent. The percentage of smokers over the age of 18 in 2015 was 15.1 percent. Find the average rate of change in the percentage of smokers over the age of 18 from 2000 to 2015.

a. -.54 percent
b. -54 percent
c. -5.4 percent
d. -15 percent

45. Triple the difference of five and a number is equal to the sum of that number and 5. What is the number?

a. 5
b. 2
c. 5.5
d. 2.5

46. In order to estimate deer population in a forest, biologists obtained a sample of deer in that forest and tagged each one of them. The sample had 300 deer in total. They returned a week later and harmlessly captured 400 deer, and found that 5 were tagged. Using this information, which of the following is the best estimate of the total number of deer in the forest?

a. 24,000 deer
b. 30,000 deer
c. 40,000 deer
d. 100,000 deer

47. What is the correct factorization of the following binomial?
$$2y^3 - 128$$
    a. $2(y + 8)(y - 8)$
    b. $2(y - 4)(y^2 + 4y + 16)$
    c. $2(y - 4)(y + 4)^2$
    d. $2(y - 4)^3$

48. What is the simplified form of $(4y^3)^4(3y^7)^2$?
    a. $12y^{26}$
    b. $2,304y^{16}$
    c. $12y^{14}$
    d. $2,304y^{26}$

49. The number of members of the House of Representatives varies directly with the total population in a state. If the state of New York has 19,800,000 residents and has 27 total representatives, how many should Ohio have with a population of 11,800,000?
    a. 10
    b. 16
    c. 11
    d. 5

50. The following set represents the test scores from a university class: {35, 79, 80, 87, 87, 90, 92, 95, 95, 98, 99}. If the outlier is removed from this set, which of the following is TRUE?
    a. The mean and the median will decrease.
    b. The mean and the median will increase.
    c. The mean and the mode will increase.
    d. The mean and the mode will decrease.

51. Which of the statements below is a statistical question?
    a. What was your grade on the last test?
    b. What were the grades of the students in your class on the last test?
    c. What kind of car do you drive?
    d. What was Sam's time in the marathon?

52. Eva Jane is practicing for an upcoming 5K run. She has recorded the following times (in minutes):
        25, 18, 23, 28, 30, 22.5, 23, 33, 20
Use the above information to answer the next three questions to the closest minute. What is Eva Jane's mean time?
    a. 26 minutes
    b. 19 minutes
    c. 25 minutes
    d. 23 minutes

53. What is the mode of Eva Jane's time?
    a. 16 minutes
    b. 20 minutes
    c. 23 minutes
    d. 33 minutes

54. What is Eva Jane's median time?
   a. 23 minutes
   b. 17 minutes
   c. 28 minutes
   d. 19 minutes

55. Use the graph below entitled "Projected Temperatures for Tomorrow's Winter Storm" to answer the question.

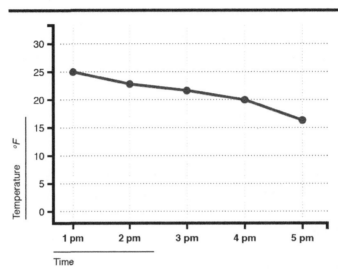

What is the expected temperature at 3:00 p.m.?
   a. 25 degrees
   b. 22 degrees
   c. 20 degrees
   d. 16 degrees

56. The function $f(x) = 3.1x + 240$ models the total U.S. population, in millions, x years after the year 1980. Use this function to answer the following question: What is the total U.S. population in 2011? Round to the nearest million.
   a. 336 people
   b. 336 million people
   c. 6,474 people
   d. 647 million people

57. What are the zeros of the following quadratic function?
$$f(x) = 2x^2 - 12x + 16$$
   a. $x = 2$ and $x = 4$
   b. $x = 8$ and $x = 2$
   c. $x = 2$ and $x = 0$
   d. $x = 0$ and $x = 4$

58. Bindee is having a barbeque on Sunday and needs 12 packets of ketchup for every 5 guests. If 60 guests are coming, how many packets of ketchup should she buy?
    a. 100
    b. 12
    c. 144
    d. 60

59. What is the volume of the cylinder below?

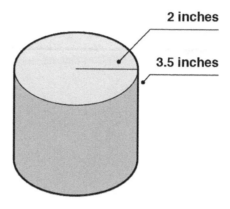

2 inches

3.5 inches

    a. 18.84 in^3
    b. 45.00 in^3
    c. 70.43 in^3
    d. 43.96 in^3

60. Given the linear function $g(x) = \frac{1}{4}x - 2$, which domain value corresponds to a range value of $\frac{1}{8}$?
    a. $\frac{17}{2}$
    b. $-\frac{63}{32}$
    c. 0
    d. $\frac{2}{17}$

61. What is the equation of the line that passes through the two points (-3, 7) and (-1, -5)?
    a. $y = 6x + 11$
    b. $y = 6x$
    c. $y = -6x - 11$
    d. $y = -6x$

62. Which of the following is the equation of a vertical line that runs through the point (1, 4)?
    a. $x = 1$
    b. $y = 1$
    c. $x = 4$
    d. $y = 4$

63. What is the area of the following figure?

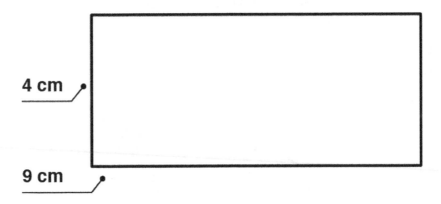

    a. 26 cm
    b. 36 cm
    c. 13 cm^2
    d. 36 cm^2

64. A jar is filled with green, yellow, and orange marbles. If $\frac{1}{4}$ of the marbles are green and $\frac{2}{7}$ are yellow, what fraction of the marbles are orange?

    a. $\frac{15}{28}$

    b. $\frac{13}{28}$

    c. $\frac{2}{3}$

    d. $\frac{3}{7}$

Answer Explanations

1. A: Surface area is a type of area, which means it is measured in square units. Cubic units are used to describe volume, which has three dimensions multiplied by one another. Quartic units describe measurements multiplied in four dimensions.

2. B: The correct choice is *B* because the definition of transposing a matrix says that the rows and columns should be switched. For example, the matrix:

$$\begin{bmatrix} 3 & 4 \\ 2 & 5 \\ 1 & 6 \end{bmatrix}$$

can be transposed into:

$$\begin{bmatrix} 3 & 2 & 1 \\ 4 & 5 & 6 \end{bmatrix}$$

Notice that the first row, 3 and 4, becomes the first column. The second row, 2 and 5, becomes the second column. This is an example of transposing a matrix.

3. A: The first step is to simplify the expression. The second term, $16^{\frac{1}{2}}$, can be rewritten as $\sqrt{16}$, since fractional exponents represent roots. This can then be simplified to 4, because 16 is the perfect square of 4. The first term can be evaluated using the power rule for exponents: an exponent to a second exponent is multiplied by that second exponent. The first term can thus be rewritten as:

$$16^{\frac{1}{4}} = \left(16^{\frac{1}{2}}\right)^{\frac{1}{2}} = 4^{\frac{1}{2}} = 2$$

Combining these terms, the expression is evaluated as follows:

$$16^{\frac{1}{4}} \times 16^{\frac{1}{2}} = 2 \times 4 = 8$$

4. D: A probability cannot be less than zero or exceed 1. Because 3/2 is equal to 1.5, this value cannot represent a probability. This value would mean something happening 150% of the time, which does not make sense.

5. B: First factor out the common factor of 25, resulting in:

$$25(25x^8 - y^4)$$

The resulting polynomial in parentheses is a difference of squares, and therefore, the entire polynomial factors into:

$$25(5x^4 + y^2)(5x^4 - y^2)$$

This can be checked by multiplication.

6. A: First, substitute -2 for x and -3 for y. This results in:

$$\frac{14(-2)}{(-3)^2} - \frac{10(-3)}{(-2)^2}$$

Multiply the numerators and evaluate the exponents in the denominators to obtain:

$$-\frac{28}{9} - \left(\frac{-30}{4}\right) = -\frac{28}{9} + \frac{15}{2}$$

The common denominator is 18, and therefore, this expression is equivalent to:

$$-\frac{56}{18} + \frac{135}{18} = \frac{79}{18}$$

7. B: First, solve for x in the given equation. Because $4^5 = 1024$, we know that $x = 5$. Then, plug 5 in for x in the given expression. Following order of operations:

$$9 \times 5^4 = 9 \times 625 = 5{,}625$$

8. A: First, multiply both sides of the equation times the quantity $(x - y)$. This results in:

$$z(x - y) = xy - 4$$

The goal is to get all y terms isolated onto one side of the equals sign. Distributing results in:

$$zx - zy = xy - 4$$

Moving the y terms to the right side and the 4 to the left side results in:

$$zx + 4 = xy + zy$$

Factoring out a y on the right side gives:

$$zx + 4 = y(x + z)$$

Finally, dividing both sides by the quantity $(x + z)$ gives:

$$y = \frac{zx + 4}{x + z}$$

9. A: Because x is directly proportional to y, $x = ky$, for some constant k. Plugging in what is given yields $8 = k\left(\frac{1}{12}\right)$, which means that $k = 96$. Therefore, $x = 96y$.

10. D: The ratio gives the proportion $\frac{x}{y} = \frac{1}{8}$. If $y = 48$, then $\frac{x}{48} = \frac{1}{8}$ means that $x = 6$. The product of x and y is therefore:

$$(6)(48) = 288$$

11. C: First, calculate the percent increase from 8 to 18 as:

$$\frac{18 - 8}{8} = 1.25 = 125\%$$

Then add 125% of 234 onto 234 to obtain:

$$292.5 + 234 = 526.5$$

12. B: Let x be equal to the fifth test score. Therefore, in order to receive, at minimum, a B in the class, the student must have:

$$\frac{74 + 76 + 82 + 84 + x}{5} = 80$$

Therefore:

$$\frac{316 + x}{5} = 80$$

Solving for x gives $316 + x = 400$, or $x = 84$. Therefore, he must receive at least an 84 out of 100 on the fifth test to receive a B in the course.

13. B: First, we multiply each score times its weight (the number of boys and the number of girls, respectively). Therefore, $3 \times 84 = 252$ and $5 \times 82 = 410$. The total weighted score is $252 + 410 = 662$. The total weight is 8 because there are 8 students. Therefore, divide 662 by 8 to obtain a class average of 82.75.

14. D: The given word has 9 letters. Therefore, in a 5-letter ordering, there are 9 choices for the first letter. Because there are no repeats allowed, there are 8 choices for the second letter, 7 choices for the third letter, 6 choices for the fourth letter, and 5 choices for the fifth letter. Multiplying these values together results in:

$$(9)(8)(7)(6)(5) = 15{,}120 \ possible \ orderings$$

15. D: There are n total cards, which means that the denominator needs to be n (the total number of outcomes). If there is only 1 odd number, then the rest are even. There are $n - 1$ even cards. Therefore, the probability of selecting an even card is $\frac{n-1}{n}$.

16. C: Because there is one $200 winner and each ticket costs $1, the probability of winning $199 is 1/1000. Also, because there is one $50 winner, the probability of winning $49 is 1/1000. The rest of the outcomes would be losing $1, which has a probability of 998/1000. The expected value is the sum of the products of each outcome and their probabilities. Therefore,

$$\text{Expected value} = \$199 \left(\frac{1}{1000}\right) + \$49 \left(\frac{1}{1000}\right) - 1 \left(\frac{998}{1000}\right)$$

$$-\frac{750}{1000} = -\$0.75$$

Expect to lose $0.75 for every ticket bought.

17. C: The smaller and larger interior angles in a parallelogram must add up to 180°. This pair is the only duo that does not have a sum of 180.

18. C: First, like terms are collected to obtain:

$$12 - 5x = -5x + 12$$

Then, if the addition principle is used to move the terms with the variable, $5x$ is added to both sides and the mathematical statement $12 = 12$ is obtained. This is always true; therefore, all real numbers satisfy the original equation.

19. D: The distributive property is used on both sides to obtain:

$$4x + 20 + 6 = 4x + 6$$

Then, like terms are collected on the left, resulting in:

$$4x + 26 = 4x + 6$$

Next, the addition principle is used to subtract $4x$ from both sides, and this results in the false statement $26 = 6$. Therefore, there is no solution.

20. A: First, the variables have to be defined. Let x be the first integer; therefore, $x + 1$ is the second integer. This is a two-step problem. The sum of three times the first and two less than the second is translated into the following expression:

$$3x + (x + 1 - 2)$$

This expression is set equal to 411 to obtain:

$$3x + (x + 1 - 2) = 411$$

The left-hand side is simplified to obtain:

$$4x - 1 = 411$$

The addition and multiplication properties are used to solve for x. First, add 1 to both sides and then divide both sides by 4 to obtain $x = 103$. The next consecutive integer is 104.

21. B: First, the information is translated into the ratio $\frac{15}{80}$. To find the percentage, translate this fraction into a decimal by dividing 15 by 80. The corresponding decimal is 0.1875. Move the decimal point two places to the right to obtain the percentage 18.75%.

22. B: If sales tax is 7.25%, the price of the car must be multiplied by 1.0725 to account for the additional sales tax. Therefore:

$$15,395 \times 1.0725 = 16,511.1375$$

This amount is rounded to the nearest cent, which is $16,511.14.

23. A: The chart is a bar chart showing how many men and women prefer each genre of movies. The dark gray bars represent the number of women, while the light gray bars represent the number of men. The light gray bars are higher and represent more men than women for the genres of Comedy and Action.

24. B: A line graph represents continuous change over time. The line on the graph is continuous and not broken, as on a scatter plot. A bar graph may show change but isn't necessarily continuous over time. A pie graph is better for representing percentages of a whole. Histograms are best used in grouping sets of data in bins to show the frequency of a certain variable.

25. C: The mean for the number of visitors during the first 4 hours is 14. The mean is found by calculating the average for the four hours. Adding up the total number of visitors during those hours gives:

$$12 + 10 + 18 + 16 = 56$$

Then:

$$56 \div 4 = 14$$

26. C: The mode for a set of data is the value that occurs the most. The grade that appears the most is 95. It's the only value that repeats in the set.

27. B: The relationship between age and time for attention span is a positive correlation because the general trend for the data is up and to the right. As age increases, so does attention span.

28. C: There are 0.006 kiloliters in 6 liters because 1 liter is 0.001 kiloliters. The conversion comes from the chart where the prefix kilo- is found three places to the left of the base unit.

29. B: The only relation in which every x-value corresponds to exactly one y-value is the relation given in Choice B, making it a function. The other relations have the same first component paired up to different second components, which goes against the definition of a function.

30. A: To find a function value, plug in the number given for the variable and evaluate the expression, using the order of operations (parentheses, exponents, multiplication, division, addition, subtraction). The function given is a polynomial function:

$$f(5) = 5^2 - 2 \times 5 + 1$$

$$f(5) = 25 - 10 + 1 = 16$$

31. C: The function given is a polynomial function. Anything can be plugged into a polynomial function to get an output. Therefore, its domain is all real numbers, which is expressed in interval notation as $(-\infty, \infty)$.

32. D: This is a parabola that opens up, as the coefficient on the x^2 term is positive. The smallest number in its range occurs when plugging zero into the function $f(0) = 5$. Any other output is a number larger than 5, even when a positive number is plugged in. When a negative number gets plugged into the function, the output is positive, and same with a positive number. Therefore, the domain is written as $[5, \infty)$ in interval notation.

33. C: First, set the functions equal to one another, resulting in:

$$x^2 + 3x + 2 = 4x + 4$$

This is a quadratic equation, so the equivalent equation in standard form is:

$$x^2 - x - 2 = 0$$

This equation can be solved by factoring into:

$$(x - 2)(x + 1) = 0$$

Setting both factors equal to zero results in $x = 2$ and $x = -1$.

34. D: There will be no more coyotes when the population is zero, so set y equal to zero and solve the quadratic equation:

$$0 = -(x - 2)^2 + 1600$$

Subtract 1600 from both sides, and divide through by -1. This results in:

$$1600 = (x - 2)^2$$

Then, take the square root of both sides. This process results in the following equation:

$$\pm 40 = x - 2$$

Adding 2 to both sides results in two solutions: $x = 42$ and $x = -38$. Because the problem involves years after 2000, the only solution that makes sense is 42. Add 42 to 2000; therefore, in 2042 there will be no more coyotes.

35. B: The ball is back at the starting point when the function is equal to 800 feet. Therefore, this results in solving the equation:

$$800 = -32t^2 + 90t + 800$$

Subtract 800 off of both sides and factor the remaining terms to obtain:

$$0 = 2t(-16t + 45)$$

Setting both factors equal to zero result in $t = 0$, which is when the ball was thrown up initially, and:

$$t = \frac{45}{16} = 2.8 \text{ seconds}$$

Therefore, it will take the ball 2.8 seconds to come back down to its staring point.

36. B: Given a rational function, the expression in the denominator can never be equal to zero. To find the domain, set the denominator equal to zero and solve for x. This results in $2 - x = 0$, and its solution is $x = 2$. This value needs to be excluded from the set of all real numbers, and therefore the domain written in interval notation is $(-\infty, 2) \cup (2, \infty)$.

37. C: The Pythagorean Theorem can be used to find the missing length x because it is a right triangle. The theorem states that $6^2 + 8^2 = x^2$, which simplifies into $100 = x^2$. Taking the positive square root of both sides results in the missing value $x = 10$.

38. A: The probability of 0.9 is closer to 1 than any of the other answers. The closer a probability is to 1, the greater the likelihood that the event will occur. The probability of 0.05 shows that it is very unlikely that an adult driver will wear their seatbelt because it is close to zero. A zero probability means that it will not occur. The probability of 0.25 is closer to zero than to one, so it shows that it is unlikely an adult will wear their seatbelt.

39. B: The goal is to first isolate the variable. The fractions can easily be cleared by multiplying the entire inequality by 5, resulting in $35 - 4x < 3$. Then, subtract 35 from both sides and divide by -4. This results in $x > 8$. Notice the inequality symbol has been flipped because both sides were divided by a negative number. The solution set, all real numbers greater than 8, is written in interval notation as $(8, \infty)$. A parenthesis shows that 8 is not included in the solution set.

40. D: This system can be solved using the method of substitution. Solving the first equation for y results in $y = 14 - 2x$. Plugging this into the second equation gives $4x + 2(14 - 2x) = -28$, which simplifies to $28 = -28$, an untrue statement. Therefore, this system has no solution because no x value will satisfy the system.

41. B: The slopes of perpendicular lines are negative reciprocals, meaning their product is equal to -1. The slope of the line given needs to be found. Its equivalent form in slope-intercept form is $y = -\frac{4}{7}x + \frac{23}{7}$, so its slope is $-\frac{4}{7}$. The negative reciprocal of this number is $\frac{7}{4}$. The only line in the options given with this same slope is:

$$y = \frac{7}{4}x - 12$$

42. C: This system can be solved using substitution. Plug the second equation in for y in the first equation to obtain $2x - 8x = 6$, which simplifies to $-6x = 6$. Divide both sides by 6 to get $x = -1$, which is then back-substituted into either original equation to obtain $y = -8$.

43. A: Division can be used to solve this problem. The division necessary is:

$$\frac{5.972 \times 10^{24}}{7.348 \times 10^{22}}$$

To compute this division, divide the constants first then use algebraic laws of exponents to divide the exponential expression.

This results in about 0.8127×10^2, which written in scientific notation is 8.127×10^1.

44. A: The formula for the rate of change is the same as slope: change in y over change in x. The y-value in this case is percentage of smokers and the x-value is year. The change in percentage of smokers from 2000 to 2015 was 8.1 percent. The change in x was $2000 - 2015 = -15$. Therefore:

$$\frac{8.1\%}{-15} = -0.54\%$$

The percentage of smokers decreased 0.54 percent each year.

45. D: Let x be the unknown number. The difference indicates subtraction, and sum represents addition. To triple the difference, it is multiplied by 3. The problem can be expressed as the following equation:

$$3(5 - x) = x + 5$$

Distributing the 3 results in:

$$15 - 3x = x + 5$$

Subtract 5 from both sides, add $3x$ to both sides, and then divide both sides by 4. This results in:

$$x = \frac{10}{4} = \frac{5}{2} = 2.5$$

46. A: A proportion should be used to solve this problem. The ratio of tagged to total deer in each instance is set equal, and the unknown quantity is a variable x. The proportion is:

$$\frac{300}{x} = \frac{5}{400}$$

Cross-multiplying gives $120,000 = 5x$, and dividing through by 5 results in 24,000.

47. B: First, the common factor 2 can be factored out of both terms, resulting in:

$$2(y^3 - 64)$$

The resulting binomial is a difference of cubes that can be factored using the rule:

$$a^3 - b^3 = (a - b)(a^2 + ab + b^2)$$

with $a = y$ and $b = 4$. Therefore, the result is:

$$2(y - 4)(y^2 + 4y + 16)$$

48. D: The exponential rules $(ab)^m = a^m b^m$ and $(a^m)^n = a^{mn}$ can be used to rewrite the expression as:

$$4^4 y^{12} \times 3^2 y^{14}$$

The coefficients are multiplied together and the exponential rule $a^m a^n = a^{m+n}$ is then used to obtain the simplified form $2,304 y^{26}$.

49. B: The number of representatives varies directly with the population, so the equation necessary is $N = k \times P$, where N is number of representatives, k is the variation constant, and P is total population in millions. Plugging in the information for New York allows k to be solved for. This process gives $27 = k \times 19.8$, so $k = 1.36$. Therefore, the formula for number of representatives given total population in millions is $N = 1.36 \times P$. Plugging in $P = 11.8$ for Ohio results in $N = 16.05$, which rounds to 16 total representatives.

50. B: The outlier is 35. When a small outlier is removed from a data set, the mean and the median increase. The first step in this process is to identify the outlier, which is the number that lies away from the given set. Once the outlier is identified, the mean and median can be recalculated. The mean will be affected because it averages all of the numbers. The median will be affected because it finds the middle number, which is subject to change because a number is lost. The mode will most likely not change because it is the number that occurs the most, which will not be the outlier if there is only one outlier.

51. B: This is a statistical question because to determine this answer one would need to collect data from each person in the class and it is expected that the answers would vary. The other answers do not require data to be collected from multiple sources; therefore, the answers will not vary.

52. C: The mean is found by adding all the times together and dividing by the number of times recorded.

$$25 + 18 + 23 + 28 + 30 + 22.5 + 23 + 33 + 20 = 222.5$$

$$\frac{222.5}{9} = 24.722$$

Rounding to the nearest minute, the mean is 25 minutes.

53. C: The mode is the time from the data set that occurs most often. The number 23 occurs twice in the data set, while all others occur only once, so the mode is 23 minutes.

54. A: To find the median of a data set, you must first list the numbers from smallest to largest, and then find the number in the middle. If there are two numbers in the middle, add the two numbers in the middle together and divide by 2. Putting this list in order from smallest to greatest yields 18, 20, 22.5, 23, 23, 25, 28, 30, and 33, where 23 is the middle number, so 23 minutes is the median.

55. B: Look on the horizontal axis to find 3:00 p.m. Move up from 3:00 p.m. to reach the dot on the graph. Move horizontally to the left to the horizontal axis to between 20 and 25; the best answer choice is 22. The answer of 25 is too high above the projected time on the graph, and the answers of 20 and 16 degrees are too low.

56. B: The variable x represents the number of years after 1980. The year 2011 was 31 years after 1980, so plug 31 into the function to obtain:

$$f(31) = 3.1 \times 31 + 240 = 336.1$$

This value rounds to 336 and represents 336 million people.

57. A: The zeros of a polynomial function are the x-values where the graph crosses the x-axis, or where $y = 0$. Therefore, set $y = 0$ and solve the polynomial equation. This quadratic can be solved using factoring, as follows:

$$0 = 2x^2 - 12x + 16$$

$$0 = 2(x^2 - 6x + 8) = 2(x - 4)(x - 2)$$

Setting both factors equal to zero results in the two solutions $x = 4$ and $x = 2$, which are the zeros of the original function.

58. C: This problem involves ratios and percentages. If 12 packets are needed for every 5 people, this statement is equivalent to the ratio $\frac{12}{5}$. The unknown amount x is the number of ketchup packets needed for 60 people. The proportion $\frac{12}{5} = \frac{x}{60}$ must be solved. Cross-multiply to obtain $12 \times 60 = 5x$. Therefore, $720 = 5x$. Divide each side by 5 to obtain $x = 144$.

59. D: The volume for a cylinder is found by using the formula:

$$V = \pi r^2 h = \pi (2 \text{ in})^2 \times 3.5 \text{ in} = 43.96 \text{ in}^3$$

60. A: The range value is given, and this is the output of the function. Therefore, the function must be set equal to $\frac{1}{8}$ and solved for x. Thus, $\frac{1}{8} = \frac{1}{4}x - 2$ needs to be solved. The fractions can be cleared by multiplying by the LCD 8. This results in $1 = 2x - 16$. Add 16 to both sides and divide by 2 to obtain:

$$x = \frac{17}{2}$$

61. C: First, the slope of the line must be found. This is equal to the change in y over the change in x, given the two points. Therefore, the slope is -6. The slope and one of the points are then plugged into the slope-intercept form of a line:

$$y - y_1 = m(x - x_1)$$

This results in:

$$y - 7 = -6(x + 3)$$

The -6 is distributed and the equation is solved for y to obtain $y = -6x - 11$.

62. A: A vertical line has the same x value for any point on the line. Other points on the line would be (1, 3), (1, 5), (1, 9), etc. Mathematically, this is written as $x = 1$. A vertical line is always of the form $x = a$ for some constant a.

63. D: The area for a rectangle is found by multiplying the length by the width. The area is also measured in square units, so the correct answer is Choice *D*. The answer of 26 is the perimeter. The answer of 13 is found by adding the two dimensions instead of multiplying.

64. B: The total fraction of green and yellow marbles is:

$$\frac{1}{4} + \frac{2}{7} = \frac{7}{28} + \frac{8}{28} = \frac{15}{28}$$

To find the fraction of orange marbles, subtract 1 by $\frac{15}{28}$:

$$1 - \frac{15}{28} = \frac{13}{28}$$

Language

Punctuation and Capitalization

Conventions of Standard English Punctuation

Rules of Capitalization

It's important to review **capitalization rules** before you begin writing. The first word of any document, and of each new sentence, is capitalized. Proper nouns, like names and adjectives derived from proper nouns, should also be capitalized. Here are some examples:

- Grand Canyon
- Pacific Palisades
- Golden Gate Bridge
- Freudian slip
- Shakespearian, Spenserian, or Petrarchan sonnet
- Irish song

Some exceptions are adjectives, originally derived from proper nouns, which through time and usage are no longer capitalized, like *quixotic, herculean*, or *draconian*. Capitals draw attention to specific instances of people, places, and things. Some categories that should be capitalized include the following:

- Brand names
- Companies
- Weekdays
- Months
- Governmental divisions or agencies
- Historical eras
- Major historical events
- Holidays
- Institutions
- Famous buildings
- Ships and other manmade constructions
- Natural and manmade landmarks
- Territories
- Nicknames
- Epithets
- Organizations
- Planets
- Nationalities
- Tribes
- Religions
- Names of religious deities
- Roads
- Special occasions, like the Cannes Film Festival or the Olympic Games

Exceptions
Related to American government, capitalize the noun Congress but not the related adjective congressional. Capitalize the noun U.S. Constitution, but not the related adjective constitutional. Many experts advise leaving the adjectives federal and state in lowercase, as in federal regulations or state water board, and only capitalizing these when they are parts of official titles or names, like Federal Communications Commission or State Water Resources Control Board. While the names of the other planets in the solar system are capitalized as names, Earth is more often capitalized only when being described specifically as a planet, like Earth's orbit, but lowercase otherwise since it is used not only as a proper noun but also to mean *land, ground, soil,* etc.

Names of animal species or breeds are not capitalized unless they include a proper noun. Then, only the proper noun is capitalized. Antelope, black bear, and yellow-bellied sapsucker are not capitalized. However, Bengal tiger, German shepherd, Australian shepherd, French poodle, and Russian blue cat are capitalized.

Other than planets, celestial bodies like the sun, moon, and stars are not capitalized. Medical conditions like tuberculosis or diabetes are lowercase; again, exceptions are proper nouns, like Epstein-Barr syndrome, Alzheimer's disease, and Down syndrome. Seasons and related terms like winter solstice or autumnal equinox are lowercase. Plants, including fruits and vegetables, like poinsettia, celery, or avocados, are not capitalized unless they include proper names, like Douglas fir, Jerusalem artichoke, Damson plums, or Golden Delicious apples.

Titles and Names
When official titles precede names, they should be capitalized, except when there is a comma between the title and name. But if a title follows or replaces a name, it should not be capitalized. For example, "the president" without a name is not capitalized, as in "The president addressed Congress." But with a name it is capitalized, like "President Obama addressed Congress." Or, "Chair of the Board Janet Yellen was appointed by President Obama." One exception is that some publishers and writers nevertheless capitalize President, Queen, Pope, etc., when these are not accompanied by names to show respect for these high offices. However, many writers in America object to this practice for violating democratic principles of equality. Occupations before full names are not capitalized, like owner Mark Cuban, director Martin Scorsese, or coach Roger McDowell.

Some universal rules for capitalization in composition titles include capitalizing the following:

- The first and last words of the title
- Forms of the verb *to be* and all other verbs
- Pronouns
- The word *not*

Universal rules for NOT capitalizing in titles include the articles *the, a,* or *an;* the conjunctions *and, or,* or *nor,* and the preposition *to,* or *to* as part of the infinitive form of a verb. The exception to all of these is UNLESS any of them is the first or last word in the title, in which case they are capitalized. Other words are subject to differences of opinion and differences among various stylebooks or methods. These include *as, but, if,* and *or,* which some capitalize and others do not. Some authorities say no preposition should ever be capitalized; some say prepositions five or more letters long should be capitalized. The *Associated Press Stylebook* advises capitalizing prepositions longer than three letters (like *about, across,* or *with*).

Ellipses

Ellipses (. . .) signal omitted text when quoting. Some writers also use them to show a thought trailing off, but this should not be overused outside of dialogue. An example of an ellipsis would be if someone is quoting a phrase out of a professional source but wants to omit part of the phrase that isn't needed: "Dr. Skim's analysis of pollen inside the body is clearly a myth . . . that speaks to the environmental guilt of our society."

Commas

Commas separate words or phrases in a series of three or more. The Oxford comma is the last comma in a series. Many people omit this last comma, but many times it causes confusion. Here is an example:

I love my sisters, the Queen of England and Madonna.

This example without the comma implies that the "Queen of England and Madonna" are the speaker's sisters. However, if the speaker was trying to say that they love their sisters, the Queen of England, as well as Madonna, there should be a comma after "Queen of England" to signify this.

Commas also separate two coordinate adjectives ("big, heavy dog") but not cumulative ones, which should be arranged in a particular order for them to make sense ("beautiful ancient ruins").

A comma ends the first of two independent clauses connected by conjunctions. Here is an example:

I ate a bowl of tomato soup, and I was hungry very shortly after.

Here are some brief rules for commas:

- Commas follow introductory words like *however, furthermore, well, why,* and *actually,* among others.

- Commas go between city and state: Houston, Texas.

- If using a comma between a surname and Jr. or Sr. or a degree like M.D., also follow the whole name with a comma: "Martin Luther King, Jr., wrote that."

- A comma follows a dependent clause beginning a sentence: "Although she was very small, . . ."

- Nonessential modifying words/phrases/clauses are enclosed by commas: "Wendy, who is Peter's sister, closed the window."

- Commas introduce or interrupt direct quotations: "She said, 'I hate him.' 'Why,' I asked, 'do you hate him?'"

Semicolons

Semicolons are used to connect two independent clauses but should never be used in the place of a comma. They can replace periods between two closely connected sentences: "Call back tomorrow; it can wait until then." When writing items in a series and one or more of them contains internal commas, separate them with semicolons, like the following:

People came from Springfield, Illinois; Alamo, Tennessee; Moscow, Idaho; and other locations.

Here are some rules concerning **hyphens**:

- Compound adjectives like state-of-the-art or off-campus are hyphenated.
- Original compound verbs and nouns are often hyphenated, like "throne-sat," "video-gamed," "no-meater."
- Adjectives ending in –ly are often hyphenated, like "family-owned" or "friendly-looking."
- "Five years old" is not hyphenated, but singular ages like "five-year-old" are.
- Hyphens can clarify. For example, in "stolen vehicle report," "stolen-vehicle report" clarifies that "stolen" modifies "vehicle," not "report."
- Compound numbers twenty-one through ninety-nine are spelled with hyphens.
- Prefixes before proper nouns/adjectives are hyphenated, like "mid-September" and "trans-Pacific."

Parentheses

Parentheses enclose information such as an aside or more clarifying information: "She ultimately replied (after deliberating for an hour) that she was undecided." They are also used to insert short, in-text definitions or acronyms: "His FBS (fasting blood sugar) was higher than normal." When parenthetical information ends the sentence, the period follows the parentheses: "We received new funds ($25,000)." Only put periods within parentheses if the whole sentence is inside them: "Look at this. (You'll be astonished.)" However, this can also be acceptable as a clause: "Look at this (you'll be astonished)." Although parentheses appear to be part of the sentence subject, they are not, and do not change subject-verb agreement: "Will (and his dog) was there."

Quotation Marks

Quotation marks are typically used when someone is quoting a direct word or phrase someone else writes or says. Additionally, quotation marks should be used for the titles of poems, short stories, songs, articles, chapters, and other shorter works. When quotations include punctuation, periods and commas should *always* be placed inside of the quotation marks.

When a quotation contains another quotation inside of it, the outer quotation should be enclosed in double quotation marks and the inner quotation should be enclosed in single quotation marks. For example: "Timmy was begging, 'Don't go! Don't leave!'" When using both double and single quotation marks, writers will find that many word-processing programs may automatically insert enough space between the single and double quotation marks to be visible for clearer reading. But if this is not the case, the writer should write/type them with enough space between to keep them from looking like three single quotation marks. Additionally, non-standard usages, terms used in an unusual fashion, and technical terms are often clarified by quotation marks. Here are some examples:

My "friend," Dr. Sims, has been micromanaging me again.

This way of extracting oil has been dubbed "fracking."

Apostrophes

One use of the **apostrophe** is followed by an *s* to indicate possession, like *Mrs. White's home* or *our neighbor's dog*. When using the *'s* after names or nouns that also end in the letter *s*, no single rule applies: some experts advise adding both the apostrophe and the *s*, like "the Jones's house," while others prefer using only the apostrophe and omitting the additional *s*, like "the Jones' house." The wisest expert advice is to pick one formula or the other and then apply it consistently. Newspapers and magazines often use *'s* after common nouns ending with *s*, but add only the apostrophe after proper nouns or names ending with *s*. One common error is to place the apostrophe before a name's final *s* instead of after it: "Ms. Hasting's book" is incorrect if the name is Ms. Hastings.

Plural nouns should not include apostrophes (e.g. "apostrophe's"). Exceptions are to clarify atypical plurals, like verbs used as nouns: "These are the do's and don'ts." Irregular plurals that do not end in *s* always take apostrophe-*s*, not *s*-apostrophe—a common error, as in "childrens' toys," which should be "children's toys." Compound nouns like mother-in-law, when they are singular and possessive, are followed by apostrophe-*s*, like "your mother-in-law's coat." When a compound noun is plural and possessive, the plural is formed before the apostrophe-*s*, like "your sisters-in-laws' coats." When two people named possess the same thing, use apostrophe-*s* after the second name only, like "Dennis and Pam's house."

Usage

Understanding the Convention of Standard English

Possessives
Possessive forms indicate possession, i.e. that something belongs to or is owned by someone or something. As such, the most common parts of speech to be used in possessive form are adjectives, nouns, and pronouns. The rule for correctly spelling/punctuating possessive nouns and proper nouns is with - *'s*, like "the woman's briefcase" or "Frank's hat." With possessive adjectives, however, apostrophes are not used: these include *my, your, his, her, its, our,* and *their*, like "my book," "your friend," "his car," "her house," "its contents," "our family," or "their property." Possessive pronouns include *mine, yours, his, hers, its, ours,* and *theirs*. These also have no apostrophes. The difference is that possessive adjectives take direct objects, whereas possessive pronouns replace them. For example, instead of using two possessive adjectives in a row, as in "I forgot my book, so Blanca let me use her book," which reads monotonously, replacing the second one with a possessive pronoun reads better: "I forgot my book, so Blanca let me use hers."

Verbs
A **verb** is a word or phrase that expresses action, feeling, or state of being. Verbs explain what their subject is *doing*. Three different types of verbs used in a sentence are action verbs, linking verbs, and helping verbs.

Action verbs show a physical or mental action. Some examples of action verbs are *play, type, jump, write, examine, study, invent, develop,* and *taste*. The following example uses an action verb:

> Kat *imagines* that she is a mermaid in the ocean.

The verb *imagines* explains what Kat is doing: she is imagining being a mermaid.

Linking verbs connect the subject to the predicate without expressing an action. The following sentence shows an example of a linking verb:

> The mango *tastes* sweet.

The verb *tastes* is a linking verb. The mango doesn't *do* the tasting, but the word *taste* links the mango to its predicate, sweet. Most linking verbs can also be used as action verbs, such as *smell, taste, look, seem, grow,* and *sound*. Saying something *is* something else is also an example of a linking verb. For example, if we were to say, "Peaches is a dog," the verb *is* would be a linking verb in this sentence, since it links the subject to its predicate.

Helping verbs are verbs that help the main verb in a sentence. Examples of helping verbs are *be, am, is, was, have, has, do, did, can, could, may, might, should,* and *must,* among others. The following are examples of helping verbs:

Jessica *is* planning a trip to Hawaii.

Brenda *does* not like camping.

Xavier *should* go to the dance tonight.

Notice that after each of these helping verbs is the main verb of the sentence: *planning, like,* and *go.* Helping verbs usually show an aspect of time.

Subject-Verb Agreement
Lack of **subject-verb agreement** is a very common grammatical error. One of the most common instances is when people use a series of nouns as a compound subject with a singular instead of a plural verb. Here is an example:

Identifying the best books, locating the sellers with the lowest prices, and paying for them *is* difficult.

The sentence should say "*are* difficult." Additionally, when a sentence subject is compound, the verb is plural:

He and his cousins *were* at the reunion.

However, if the conjunction connecting two or more singular nouns or pronouns is "or" or "nor," the verb must be singular to agree:

That pen or another one like it is in the desk drawer.

If a compound subject includes both a singular noun and a plural one, and they are connected by "or" or "nor," the verb must agree with the subject closest to the verb: "Sally or her sisters go jogging daily"; but "Her sisters or Sally goes jogging daily."

Simply put, singular subjects require singular verbs and plural subjects require plural verbs. A common source of agreement errors is not identifying the sentence subject correctly. For example, people often write sentences incorrectly like, "The group of students *were* complaining about the test." The subject is not the plural "students" but the singular "group." Therefore, the correct sentence should read, "The group of students *was* complaining about the test." The converse also applies, for example, in this incorrect sentence: "The facts in that complicated court case *is* open to question." The subject of the sentence is not the singular "case" but the plural "facts." Hence the sentence would correctly be written: "The facts in that complicated court case *are* open to question." New writers should not be misled by the distance between the subject and verb, especially when another noun with a different number intervenes as in these examples. The verb must agree with the subject, not the noun closest to it.

Inappropriate Shifts in Verb Tense

Verb tense helps indicate when an action or a state occurred or existed.

Simple present tense is used to indicate that the action or state of being is currently happening or happens regularly:

> He *plays* guitar.

Present continuous tense is used to indicate that the action or state of being is in progress. It is formed by the proper to be + verb + *-ing*.

> Unfortunately, I can't go to the park right now. I *am fixing* my bicycle.

Past tense is used to indicate that the action or state of being occurred previously. It should be noted, however, that in conversational English, speakers frequently use a mix of present and past tense, or simply present tense when describing events in the past. With that said, it is important for writers (and speakers in formal situations) to be consistent and grammatically correct in their verb tenses to avoid confusing readers. Consider the following passage:

> I scored a goal in our soccer game last Saturday. At the start of the first half, Billy kicked me the ball. I run toward it and strike it directly toward the goal. It goes in and we won the game!

The passage above inappropriately switches from past tense—*scored, kicked*—to present tense—*run, strike, goes*—and then back to past tense—*won*. Instead, past tense should be carried throughout the passage:

> I *scored* a goal in our soccer game last Saturday. At the start of the first half, Billy *kicked* me the ball. I *ran* toward it and strike it directly toward the goal. It *went* in and we *won* the game!

Noun-Noun Agreement

When multiple nouns are included in the same sentence and are related to one another in that sentence, they need to agree in number. This means that if one noun is singular, all other related nouns in the sentence must be singular as well. Similarly, if one noun is plural, the rest should follow in form and be plural as well. Consider the following sentence with an error in noun-noun agreement:

> Mary and Sharon both have jobs as a teacher.

Because the noun *jobs* is plural, the noun *teachers*, which is also plural, must be used in place of *teacher*, which is singular.

> Mary and Sharon both have jobs as teachers.

Pronouns

There are three **pronoun** cases: subjective case, objective case, and possessive case. Pronouns as subjects are pronouns that replace the subject of the sentence, such as *I, you, he, she, it, we, they* and *who*. Pronouns as objects replace the object of the sentence, such as *me, you, him, her, it, us, them,* and *whom*. Pronouns that show possession are *mine, yours, hers, its, ours, theirs,* and *whose*. The following are examples of different pronoun cases:

- Subject pronoun: *She* ate the cake for her birthday. *I* saw the movie.
- Object pronoun: You gave *me* the card last weekend. She gave the picture to *him*.
- Possessive pronoun: That bracelet you found yesterday is *mine*. *His* name was Casey.

Pronoun-Antecedent Agreement

Pronouns within a sentence must refer specifically to one noun, known as the **antecedent**. Sometimes, if there are multiple nouns within a sentence, it may be difficult to ascertain which noun belongs to the pronoun. It's important that the pronouns always clearly reference the nouns in the sentence so as not to confuse the reader. Here's an example of an unclear pronoun reference:

> After Catherine cut Libby's hair, David bought her some lunch.

The pronoun in the examples above is *her*. The pronoun could either be referring to *Catherine* or *Libby*. Here are some ways to write the above sentence with a clear pronoun reference:

> After Catherine cut Libby's hair, David bought Libby some lunch.

> David bought Libby some lunch after Catherine cut Libby's hair.

But many times, the pronoun will clearly refer to its antecedent, like the following:

> After David cut Catherine's hair, he bought her some lunch.

Intensive Pronoun Errors

An **intensive pronoun** ends in "self" or "selves" and adds emphasis to the sentence's subject or antecedent. Like **reflexive pronouns**, intensive pronouns include the singular pronouns *myself, yourself, himself, herself,* and *itself,* and the plural pronouns *ourselves, yourselves,* and *themselves*. However, intensive and reflexive pronouns differ in that removing a reflexive pronoun will cause the sentence to no longer make sense, whereas intensive pronouns can be removed because they only add emphasis; they are not mandatory. An example of a sentence with an intensive pronoun is the following:

> We want to hear the author herself read the story.

The intensive pronoun *herself* adds emphasis that the speakers want to specifically hear the author read the story, rather than anyone else.

The most common error in the use of intensive pronouns is choosing the wrong pronoun; for example, using the plural pronoun when the singular one is needed, or using a singular pronoun when a plural one is needed. Using the same example from above, an error in agreement occurs in the following sentence:

> We want to hear the author themselves read the story.

Author is singular and *themselves* is plural, so there is an error in number agreement.

Pronoun Number and Person Errors

Pronouns must agree in number and person. However, it is common, unfortunately, for writers to shift between persons or numbers when using pronouns. For example, a sentence might start with third person pronouns (*he, she, it, they,* etc.), but then switch to second person. Or, a sentence might start in second person and switch to first or third person. The following sentence contains a pronoun shift in person:

> If you drink more water, most people see improvements in their skin and body composition.

This example begins with second person (using the pronoun *you*), but switches to third person (*their*). Consistency in person is needed. Therefore, the sentence should be one of the following two options:

> If you drink more water, you will likely see improvements in your skin and body composition.

> If they drink more water, most people see improvements in their skin and body composition.

Inappropriate pronoun shifts also occur when writers switch from using singular pronouns to plural ones, or vice versa. Sometimes, sentences will have errors in pronoun number and person.

> Everyone should keep a journal about their life because you will want to pass the stories of your life along.

These sentences are usually easier to spot because the errors are two-fold and more apparent.

Adjectives

Adjectives are descriptive words that modify nouns or pronouns. They may occur before or after the nouns or pronouns they modify in sentences. For example, in "This is a big house," *big* is an adjective modifying or describing the noun *house*. In "This house is big," the adjective is at the end of the sentence rather than preceding the noun it modifies.

A rule of punctuation that applies to adjectives is to separate a series of adjectives with commas. For example, "Their home was a large, rambling, old, white, two-story house." A comma should never separate the last adjective from the noun, though.

Adverbs

Whereas adjectives modify and describe nouns or pronouns, **adverbs** modify and describe adjectives, verbs, or other adverbs. Adverbs can be thought of as answers to questions in that they describe when, where, how, how often, how much, or to what extent.

Many (but not all) adjectives can be converted to adverbs by adding *–ly*. For example, in "She is a quick learner," *quick* is an adjective modifying *learner*. In "She learns quickly," *quickly* is an adverb modifying *learns*. One exception is *fast*. *Fast* is an adjective in "She is a fast learner." However, *–ly* is never added to the word *fast*—it retains the same form as an adverb in "She learns fast."

Transitional Words and Phrases

In connected writing, some sentences naturally lead to others, whereas in other cases, a new sentence expresses a new idea. We use **transitional phrases** to connect sentences and the ideas they convey. This makes the writing coherent. Transitional language also guides the reader from one thought to the next. For example, when pointing out an objection to the previous idea, starting a sentence with "However," "But," or "On the other hand" is transitional. When adding another idea or detail, writers use "Also," "In addition," "Furthermore," "Further," "Moreover," "Not only," etc. Readers have difficulty perceiving connections between ideas without such transitional wording.

Spelling

Conventions of Standard English Spelling

Homonyms and Homographs

Homophones are words that sound the same in speech but have different spellings and meanings. For example, *to, too,* and *two* all sound alike, but have three different spellings and meanings. Homophones with different spellings are also called **heterographs**. **Homographs** are words that are spelled identically but have different meanings. If they also have different pronunciations, they are **heteronyms**. For instance, *tear* pronounced one way means a drop of liquid formed by the eye; pronounced another way, it means to rip. Homophones that are also homographs are **homonyms**. For example, *bark* can mean the outside of a tree or a dog's vocalization; both meanings have the same spelling. *Stalk* can mean a plant stem or to pursue and/or harass somebody; these are spelled and pronounced the same. *Rose* can mean a flower or the past tense of *rise*. Many non-linguists confuse things by using "homonym" to mean sets of words that are homophones but not homographs, and also those that are homographs but not homophones.

The word *row* can mean to use oars to propel a boat; a linear arrangement of objects or print; or an argument. It is pronounced the same with the first two meanings, but differently with the third. Because it is spelled identically regardless, all three meanings are homographs. However, the two meanings pronounced the same are homophones, whereas the one with the different pronunciation is a heteronym. By contrast, the word *read* means to peruse language, whereas the word *reed* refers to a marsh plant. Because these are pronounced the same way, they are homophones; because they are spelled differently, they are heterographs. Homonyms are both homophones and homographs—pronounced and spelled identically, but with different meanings. One distinction between homonyms is of those with separate, unrelated etymologies, called "true" homonyms, e.g. *skate* meaning a fish or *skate* meaning to glide over ice/water. Those with common origins are called polysemes or polysemous homonyms, e.g. the *mouth* of an animal/human or of a river.

Irregular Plurals

One type of irregular English plural involves words that are spelled the same whether they are singular or plural. These include *deer, fish, salmon, trout, sheep, moose, offspring, species, aircraft,* etc. The spelling rule for making these words plural is simple: they do not change. Another type of irregular English plurals does change from singular to plural form, but it does not take regular English *–s* or *–es* endings. Their irregular plural endings are largely derived from grammatical and spelling conventions in the other languages of their origins, like Latin, German, and vowel shifts and other linguistic mutations. Some examples of these words and their irregular plurals include *child* and *children; die* and *dice; foot* and *feet; goose* and *geese; louse* and *lice; man* and *men; mouse* and *mice; ox* and *oxen; person* and *people; tooth* and *teeth;* and *woman* and *women.*

Contractions

Contractions are formed by joining two words together, omitting one or more letters from one of the component words, and replacing the omitted words with an apostrophe. An obvious yet often forgotten rule for spelling contractions is to place the apostrophe where the letters were omitted; for example, spelling errors like *did'nt* for *didn't. Didn't* is a contraction of *did not.* Therefore, the apostrophe replaces the "o" that is omitted from the "not" component. Another common error is confusing contractions with **possessives** because both include apostrophes, e.g. spelling the possessive *its* as "it's," which is a contraction of "it is"; spelling the possessive *their* as "they're," a contraction of "they are"; spelling the possessive *whose* as "who's," a contraction of "who is"; or spelling the possessive *your* as "you're," a contraction of "you are."

Frequently Misspelled Words

One source of spelling errors is not knowing whether to drop the final letter *e* from a word when its form is changed by adding an ending to indicate the past tense or progressive participle of a verb, converting an adjective to an adverb, a noun to an adjective, etc. Some words retain the final *e* when another syllable is added; others lose it. For example, *true* becomes *truly, argue* becomes *arguing, come* becomes *coming, write* becomes *writing*, and *judge* becomes *judging*. In these examples, the final *e* is dropped before adding the ending. But *severe* becomes *severely*, *complete* becomes *completely*, *sincere* becomes *sincerely*, *argue* becomes *argued*, and *care* becomes *careful*. In these instances, the final *e* is retained before adding the ending. Note that some words, like *argue* in these examples, drops the final *e* when the –*ing* ending is added to indicate the participial form; but the regular past tense ending of –*ed* makes it *argued*, in effect, replacing the final *e* so that *arguing* is spelled without an *e* but *argued* is spelled with one.

Some English words contain the vowel combination of *ei*, while some contain the reverse combination of *ie*. Many people confuse these. Some examples include these:

> *ceiling, conceive, leisure, receive, weird, their, either, foreign, sovereign, neither, neighbors, seize, forfeit, counterfeit, height, weight, protein,* and *freight*

Words with *ie* include *piece, believe, chief, field, friend, grief, relief, mischief, siege, niece, priest, fierce, pierce, achieve, retrieve, hygiene, science,* and *diesel*. A rule that also functions as a mnemonic device is "I before E except after C, or when sounded like A as in 'neighbor' or 'weigh'." However, it is obvious from the list above that many exceptions exist.

Many people often misspell certain words by confusing whether they have the vowel *a, e,* or *i,* frequently in the middle syllable of three-syllable words or beginning the last syllables that sound the same in different words. For example, in the following correctly-spelled words, the vowel in boldface is the one people typically get wrong by substituting one or either of the others for it:

> c**e**metery, quant**i**ties, ben**e**fit, priv**i**lege, unpleas**a**nt, sep**a**rate, independ**e**nt, excell**e**nt, cat**e**gories, indispens**a**ble, and irrelev**a**nt

The words with final syllables that sound the same when spoken but are spelled differently include *unpleasant, independent, excellent,* and *irrelevant*. Another source of misspelling is whether or not to double consonants when adding suffixes. For example, we double the last consonant before –*ed* and –*ing* endings in *controlled, beginning, forgetting, admitted, occurred, referred,* and *hopping;* but we do not double the last consonant before the suffix in *shining, poured, sweating, loving, hating, smiling,* and *hoping*.

One way in which people misspell certain words frequently is by failing to include letters that are silent. Some letters are articulated when pronounced correctly but elided in some people's speech, which then transfers to their writing. Another source of misspelling is the converse: people add extraneous letters. For example, some people omit the silent *u* in *guarantee,* overlook the first *r* in *surprise,* leave out the *z* in *realize,* fail to double the *m* in *recommend,* leave out the middle *i* from *aspirin,* and exclude the *p* from *temperature*. The converse error, adding extra letters, is common in words like *until* by adding a second *l* at the end; or by inserting a superfluous syllabic *a* or *e* in the middle of *athletic,* reproducing a common mispronunciation.

Composition

Sentence Structures

Incomplete Sentences
Sentence fragments are caused by absent subjects, absent verbs, or dangling/uncompleted dependent clauses. Every sentence must have a subject and a verb to be complete. An example of a fragment is

"Raining all night long," because there is no subject present. "It was raining all night long" is one correction. Another example of a sentence fragment is the second part in "Many scientists think in unusual ways. Einstein, for instance." The second phrase is a fragment because it has no verb. One correction is "Many scientists, like Einstein, think in unusual ways." Finally, look for "cliffhanger" words like *if, when, because,* or *although* that introduce dependent clauses, which cannot stand alone without an independent clause. For example, to correct the sentence fragment "If you get home early," add an independent clause: "If you get home early, we can go dancing."

Run-On Sentences

A **run-on sentence** combines two or more complete sentences without punctuating them correctly or separating them. For example, a run-on sentence caused by a lack of punctuation is the following:

> There is a malfunction in the computer system however there is nobody available right now who knows how to troubleshoot it.

One correction is, "There is a malfunction in the computer system; however, there is nobody available right now who knows how to troubleshoot it." Another is, "There is a malfunction in the computer system. However, there is nobody available right now who knows how to troubleshoot it."

An example of a comma splice of two sentences is the following:

> Jim decided not to take the bus, he walked home.

Replacing the comma with a period or a semicolon corrects this. Commas that try and separate two independent clauses without a contraction are considered **comma splices**.

Parallel Sentence Structures

Parallel structure in a sentence matches the forms of sentence components. Any sentence containing more than one description or phrase should keep them consistent in wording and form. Readers can easily follow writers' ideas when they are written in parallel structure, making it an important element of correct sentence construction. For example, this sentence lacks parallelism: "Our coach is a skilled manager, a clever strategist, and works hard." The first two phrases are parallel, but the third is not. Correction: "Our coach is a skilled manager, a clever strategist, and a hard worker." Now all three phrases match in form. Here is another example:

> Fred intercepted the ball, escaped tacklers, and a touchdown was scored.

This is also non-parallel. Here is the sentence corrected:

> Fred intercepted the ball, escaped tacklers, and scored a touchdown.

Sentence Fluency

For fluent composition, writers must use a variety of sentence types and structures, and also ensure that they smoothly flow together when they are read. To accomplish this, they must first be able to identify fluent writing when they read it. This includes being able to distinguish among simple, compound, complex, and compound-complex sentences in text; to observe variations among sentence types, lengths, and beginnings; and to notice figurative language and understand how it augments sentence length and imparts musicality. Once students/writers recognize superior fluency, they should revise their own writing to be more readable and fluent. They must be able to apply acquired skills to revisions before being able to apply them to new drafts.

One strategy for revising writing to increase its **sentence fluency** is flipping sentences. This involves rearranging the word order in a sentence without deleting, changing, or adding any words. For example, the student or other writer who has written the sentence, "We went bicycling on Saturday" can revise it to, "On Saturday, we went bicycling." Another technique is using appositives. An **appositive** is a phrase or word that renames or identifies another adjacent word or phrase. Writers can revise for sentence fluency by inserting main phrases/words from one shorter sentence into another shorter sentence, combining them into one longer sentence, e.g. from "My cat Peanut is a gray and brown tabby. He loves hunting rats." to "My cat Peanut, a gray and brown tabby, loves hunting rats." Revisions can also connect shorter sentences by using conjunctions and commas and removing repeated words: "Scott likes eggs. Scott is allergic to eggs" becomes "Scott likes eggs, but he is allergic to them."

One technique for revising writing to increase sentence fluency is "padding" short, simple sentences by adding phrases that provide more details specifying why, how, when, and/or where something took place. For example, a writer might have these two simple sentences: "I went to the market. I purchased a cake." To revise these, the writer can add the following informative dependent and independent clauses and prepositional phrases, respectively: "Before my mother woke up, I sneaked out of the house and went to the supermarket. As a birthday surprise, I purchased a cake for her." When revising sentences to make them longer, writers must also punctuate them correctly to change them from simple sentences to compound, complex, or compound-complex sentences.

Structural Relationships
Placement of Phrases and Clauses Within a Sentence
Clauses contain a subject and a predicate, while **phrases** only contain a noun with no verb or a verb with no noun, and they do not have a predicate. Clauses can be independent or dependent. **Independent clauses** can stand on their own as simple sentences. For example:

> She collects stamps.

Dependent clauses need independent clauses to form a complete sentence; they cannot stand alone. For example:

> Although she collects stamps . . .

Phrases can take on many forms including prepositional phrases, gerund phrases, noun phrases, infinitive phrases, verb phrases, etc. Regardless of the type, phrases cannot stand alone as complete sentence.

Phrases and clauses must be appropriately placed in a sentence such that it is clear to readers what they are modifying.

Consider the following misplaced prepositional phrase:

> At the bottom of the pile, Lila found her scarf.

In the above sentence, the phrase *at the bottom of the pile* is intended to modify the noun phrase *the scarf* by providing details about where the scarf was found. However, as written, the prepositional phrase is next to the subject, Lila, so it is modifying Lila. This is incorrect because presumably Lila herself wasn't at the bottom of a pile, her scarf was.

Misplaced and Dangling Modifiers
Modifiers are optional elements that can clarify or add details about a phrase or another element of a sentence. They are a dependent phrase and removing them usually does not affect the grammatical

correctness of the sentence; however, the meaning will be changed because, as their name implies, modifiers modify another element in the sentence.

Consider the following:

Nico loves sardines.

Nico, who is three years old, loves sardines.

The first simple sentence is grammatically correct; however, we learn a lot more from the second sentence, which contains the modifier *who is three years old.* Sardines tend to be a food that young children don't like, so adding the modifier helps readers see why Nico loving sardines is noteworthy.

Beginning writers sometimes place modifiers incorrectly. Then, instead of enhancing comprehension and providing helpful description for the reader, the modifier causes more confusion. A **misplaced modifier** is located incorrectly in relation to the phrase or word it modifies. Consider the following sentence:

Because it is salty, Nico loves fish.

The modifier in this sentence is "because they are salty," and the noun it is intended to modify is "fish." However, due to the erroneous placement of the modifier next to the subject, Nico, the sentence is actually saying that Nico is salty.

Nico loves fish because it is salty.

The modifier is now adjacent to the appropriate noun, clarifying which of the two elements is salty.

Dangling modifiers are so named because they modify a phrase or word that is not clearly found in the sentence, making them rather unattached. They are not intended to modify the word or phrase they are placed next to. Consider the following:

Walking home from school, the sky opened and Bruce got drenched.

The modifier here, "walking home from school," should modify who was walking (Bruce). Instead, the noun immediately after the modifier is "the sky"—but the sky was not walking home from school. Although not always the case, dangling modifiers are often found at the beginning of a sentence.

Coordinating and Subordinating Conjunctions
Conjunctions connect or coordinate words, phrases, clauses, or sentences together, typically as a way to demonstrate a relationship.

Tony has a cat *and* a rabbit.

Tony likes animals, *but* he is afraid of snakes.

Coordinating conjunctions join words or phrases that have equal rank or emphasis. There are seven coordinating conjunctions, all short words, that can be remembered by the mnemonic FANBOYS: *for, and, nor, but, or, yet, so.* They can join two words that are of the same part of speech (two verbs, two adjectives, two adverbs, or two nouns). They can also connect two phrases or two independent clauses.

Subordinating conjunctions help transition and connect two elements in the sentence, but in a way that diminishes the importance of the one it introduces, known as the dependent, or subordinate, clause. They include words like *because, since, unless, before, after, whereas, if,* and *while.*

Fragments and Run-Ons

Every sentence must have a subject and a verb to be complete. As mentioned, **sentence fragments** are caused by absent subjects, absent verbs, or dangling/uncompleted dependent clauses. An example of a fragment is "Raining all night long," because there is no subject present. "It was raining all night long" is one correction. Another example of a sentence fragment is the second part in "Many scientists think in unusual ways. Einstein, for instance." The second phrase is a fragment because it has no verb. One correction is "Many scientists, like Einstein, think in unusual ways." Finally, look for "cliffhanger" words like *if, when, because,* or *although* that introduce dependent clauses, which cannot stand alone without an independent clause. For example, to correct the sentence fragment "If you get home early," add an independent clause: "If you get home early, we can go dancing."

Correlative Conjunctions

Correlative conjunctions are pairs of conjunctions that must both be used in the sentence, though in different spots, to make the sentence grammatically sound. They help relate one aspect of the sentence to another. Examples of correlative conjunction pairs are *neither/nor, both/and, not/but, either/or, as/as, such/that* and *rather/than.* They tend to be more like coordinating conjunctions rather than subordinating conjunctions in that they typically connect two words or phrases of equal weight in the sentence.

Parallel Structure

As mentioned, **parallel structure** in a sentence matches the forms of sentence components. Any sentence containing more than one description or phrase should keep them consistent in wording and form. Readers can easily follow writers' ideas when they are written in parallel structure, making it an important element of correct sentence construction. For example, this sentence lacks parallelism: "Our coach is a skilled manager, a clever strategist, and works hard." The first two phrases are parallel, but the third is not. Correction: "Our coach is a skilled manager, a clever strategist, and a hard worker." Now all three phrases match in form. Here is another example:

Fred intercepted the ball, escaped tacklers, and a touchdown was scored.

This is also non-parallel. Here is the sentence corrected:

Fred intercepted the ball, escaped tacklers, and scored a touchdown.

Developing a Well-Organized Paragraph

A **paragraph** is a series of connected and related sentences addressing one topic. Writing good paragraphs benefits writers by helping them to stay on target while drafting and revising their work. It benefits readers by helping them to follow the writing more easily. Regardless of how brilliant their ideas may be, writers who do not present them in organized ways will fail to engage readers—and fail to accomplish their writing goals. A fundamental rule for paragraphing is to confine each paragraph to a single idea. When writers find themselves transitioning to a new idea, they should start a new paragraph. However, a paragraph can include several pieces of evidence supporting its single idea; and it can include several points if they are all related to the overall paragraph topic. When writers find each point becoming lengthy, they may choose instead to devote a separate paragraph to every point and elaborate upon each more fully.

An effective paragraph should have these elements:

- Unity: One major discussion point or focus should occupy the whole paragraph from beginning to end.

- Coherence: For readers to understand a paragraph, it must be coherent. Two components of coherence are logical and verbal bridges. In logical bridges, the writer may write consecutive sentences with parallel structure or carry an idea over across sentences. In verbal bridges, writers may repeat key words across sentences.

- A topic sentence: The paragraph should have a sentence that generally identifies the paragraph's thesis or main idea.

- Sufficient development: To develop a paragraph, writers can use the following techniques after stating their topic sentence:

    o Define terms
    o Cite data
    o Use illustrations, anecdotes, and examples
    o Evaluate causes and effects
    o Analyze the topic
    o Explain the topic using chronological order

A **topic sentence** identifies the main idea of the paragraph. Some are explicit, some implicit. The topic sentence can appear anywhere in the paragraph. However, many experts advise beginning writers to place each paragraph topic sentence at or near the beginning of its paragraph to ensure that their readers understand what the topic of each paragraph is. Even without having written an explicit topic sentence, the writer should still be able to summarize readily what subject matter each paragraph addresses. The writer must then fully develop the topic that is introduced or identified in the topic sentence. Depending on what the writer's purpose is, they may use different methods for developing each paragraph.

Two main steps in the process of organizing paragraphs and essays should both be completed after determining the writing's main point, while the writer is planning or outlining the work. The initial step is to give an order to the topics addressed in each paragraph. Writers must have logical reasons for putting one paragraph first, another second, etc. The second step is to sequence the sentences in each paragraph. As with the first step, writers must have logical reasons for the order of sentences. Sometimes the work's main point obviously indicates a specific order.

Practice Questions

Punctuation and Capitalization

1. Choose the sentence that contains an error in punctuation or capitalization. If there are no errors, select Choice *D*.
    a. After school, we went to Denis' and Pam's house.
    b. Michael's dog, Roxy, loves to eat peanut butter treats.
    c. On Thursday, I have an appointment with a realtor to look at houses in the city.
    d. No errors.

2. Choose the sentence that contains an error in punctuation or capitalization. If there are no errors, select Choice *D*.
    a. Some of Tim's friends parked their cars along the street.
    b. All of your friends' invitations were sent the same day.
    c. Who's car is parked on the street with its lights on?
    d. No errors.

3. Choose the sentence that contains an error in punctuation or capitalization. If there are no errors, select Choice *D*.
    a. We don't go out as much because babysitters, gasoline, and parking are expensive.
    b. All Teachers should provide their students with clear expectations.
    c. You're going to be late if you don't leave soon.
    d. No errors.

4. Choose the sentence that contains an error in punctuation or capitalization. If there are no errors, select Choice *D*.
    a. Delegates attended from springfield, Illinois; alamo, Tennessee; moscow, Idaho; and other places.
    b. Benny went to the library to look for research materials for his report on famous African Americans.
    c. Every student must consult their advisor first.
    d. No errors.

5. Choose the sentence that contains an error in punctuation or capitalization. If there are no errors, select Choice *D*.
    a. The family moved to Greece after they sold their home in Duluth, Minnesota.
    b. The most important aim, according to the author, is to retain the truth of the work.
    c. Government assistance programs such as Medicaid and Children's Health Insurance Program (CHIP) provide low-cost medical benefits to low-income adults, children, seniors, pregnant women, and people with disabilities.
    d. No errors.

6. Choose the sentence that contains an error in punctuation or capitalization. If there are no errors, select Choice *D*.
    a. Please give the fruitcake to Shirley and me.
    b. Sandy will have finished by the end of the second semester.
    c. Arnold and Ginger worked together like a well-oiled machine.
    d. No errors.

7. Choose the sentence that contains an error in punctuation or capitalization. If there are no errors, select Choice *D*.

    a. A new shipment of twelve boxes has been delivered into your unit, and they're stacked up in the supply room.

    b. The boys never met, they were strangers.

    c. You tell your coworker you're sorry, but you don't think it's right that your team should suffer for someone else's mistake.

    d. No errors.

8. Choose the sentence that contains an error in punctuation or capitalization. If there are no errors, select Choice *D*.

    a. 'I have to call my mom, because she said, "Remind me to pick up milk on my way home."'

    b. "I don't think he will attend, because he said, 'I am extremely busy.'"

    c. "Julius wants to go tomorrow, because he said, 'It's less expensive.'"

    d. No errors.

9. Choose the sentence that contains an error in punctuation or capitalization. If there are no errors, select Choice *D*.

    a. Tanya's goal is to make it to the Olympic Games.

    b. Ms. Lamdin's class is studying the Constitution and constitutional rights.

    c. In the Middle Ages, many people died from the plague.

    d. No errors.

10. Choose the sentence that contains an error in punctuation or capitalization. If there are no errors, select Choice *D*.

    a. The governor is speaking at a political rally on Saturday.

    b. The term length for the President is four years.

    c. One of the most recognizable figures among Americans is the President of the United States.

    d. No errors.

11. Choose the sentence that contains an error in punctuation or capitalization. If there are no errors, select Choice *D*.

    a. It's not easy to get my tuba onto the bus with me.

    b. Bertha's boss bought everyone in the office Chinese food for lunch.

    c. Is that type of granola on sale this week.

    d. No errors.

12. Choose the sentence that contains an error in punctuation or capitalization. If there are no errors, select Choice *D*.

    a. My favorite cookies are homemade; packaged cookies just don't have that special quality that fresh ones do.

    b. Jim bought several things at the market: bread, cheese, laundry detergent, and canned soup.

    c. After church, Wang bought a birthday card for his friend Selma.

    d. No errors.

Usage

13. Choose the sentence that contains an error in usage. If there are no errors, select Choice *D*.
    a. The book has a crack in its spine.
    b. Martha asked, "Will you tuck me in?"
    c. The turbulence effected my ability to read on the plane.
    d. No errors.

14. Choose the sentence that contains an error in usage. If there are no errors, select Choice *D*.
    a. Her too-year-old sister is deaf.
    b. My favorite types of music, jazz and classical, are not very popular amongst teenagers.
    c. Before you ask, let me assure you, I already picked up the mess.
    d. No errors.

15. Choose the sentence that contains an error in usage. If there are no errors, select Choice *D*.
    a. The moon is full tonight.
    b. Our favorite pastime is making homemade jam.
    c. Can I get you anymore soda?
    d. No errors.

16. Choose the sentence that contains an error in usage. If there are no errors, select Choice *D*.
    a. According to Sam, our report isn't due until Friday.
    b. I forgot to feed the parking meter, and I got slammed with a ticket!
    c. Are your ears pierced or are those clip-on earrings?
    d. No errors.

17. Choose the sentence that contains an error in usage. If there are no errors, select Choice *D*.
    a. The bunny had long, fluffy ears and a bushy tail.
    b. Please call Sharon and I after you get home from your trip.
    c. Don't go there alone; it's dangerous.
    d. No errors.

18. Choose the sentence that contains an error in usage. If there are no errors, select Choice *D*.
    a. We saw three mooses on our camping trip.
    b. My little sister likes dancing, singing, and acting.
    c. The bird's song sounded cheerful and energetic.
    d. No errors.

19. Choose the sentence that contains an error in usage. If there are no errors, select Choice *D*.
    a. Its difficult to study with the television blaring in the background.
    b. After the seminar, we had a test.
    c. "Do you want more soup?" Tucker asked.
    d. No errors.

20. Choose the sentence that contains an error in usage. If there are no errors, select Choice *D*.
    a. The school bus broke down on the side of the road.
    b. My aunt took us to the Indian restaurant; we got the most delicious food.
    c. Meryl can't afford her prescriptions this month because her insurance will not cover them.
    d. No errors.

21. Choose the sentence that contains an error in usage. If there are no errors, select Choice *D*.
    a. It's not everyday that you find a four leaf clover!
    b. Trish raised her hand and said, "The answer is Austin, Texas."
    c. I wish I didn't watch so much TV, but it's hard to fight through depression.
    d. No errors.

22. Choose the sentence that contains an error in usage. If there are no errors, select Choice *D*.
    a. The kite's tale got caught in the tree.
    b. The lady at the supermarket bought treats and toys for her puppy.
    c. No one said that school would be easy.
    d. No errors.

23. Choose the sentence that contains an error in usage. If there are no errors, select Choice *D*.
    a. Pistachios are my favorite kind of nut, although they're expensive.
    b. Having finished the soup, the bowl was washed by Walter.
    c. I usually wear an apron when I cook to protect my clothing.
    d. No errors.

24. Choose the sentence that contains an error in usage. If there are no errors, select Choice *D*.
    a. On Tuesday, Miranda has to go to the post office for stamps.
    b. The soccer team lost the game; their star player was out sick.
    c. It rained so hard that we stomped, jumped, and splashed in the puddles.
    d. No errors.

25. Choose the sentence that contains an error in usage. If there are no errors, select Choice *D*.
    a. Our hopes was dashed after the agent told us the tickets were sold out.
    b. She directed her snide remarks at no one in particular.
    c. Janis forget several key ingredients in the chicken casserole.
    d. No errors.

26. Choose the sentence that contains an error in usage. If there are no errors, select Choice *D*.
    a. Moreover, your library card has expired.
    b. I don't have any fears accept sharks.
    c. It will be difficult to reach me once I've left the office tonight.
    d. No errors.

27. Choose the sentence that contains an error in usage. If there are no errors, select Choice *D*.
    a. It's going to be a late night.
    b. The baby cried and cried all night.
    c. Their was no way to fix the problem.
    d. No errors.

28. Choose the sentence that contains an error in usage. If there are no errors, select Choice *D*.
    a. Are there any additional problems plaguing you?
    b. Uncle Theo has to follow a gluten-free diet.
    c. "May I help you, Ted?"
    d. No errors.

29. Choose the sentence that contains an error in usage. If there are no errors, select Choice *D*.
    a. Avocados are so popular these days, though I despise them.
    b. My favorite hand soap smells like gingerbread cookies.
    c. She likes hiking, dancing, and to ride her bike.
    d. No errors.

30. Choose the sentence that contains an error in usage. If there are no errors, select Choice *D*.
    a. Watching basketball is more fun then watching baseball.
    b. Since Cathy forgot her homework, she lost her chance at winning the academic award.
    c. Let's order extra pizza this week because we always run out.
    d. No errors.

31. Choose the sentence that contains an error in usage. If there are no errors, select Choice *D*.
    a. I have been on the phone with tech support for fifty minutes.
    b. What Greg didn't expect was a raise from his boss.
    c. The bread at the grocery store was stale.
    d. No errors.

32. Choose the sentence that contains an error in usage. If there are no errors, select Choice *D*.
    a. "Please, Alice, stop interrupting me."
    b. We needed something to do, so we ran laps.
    c. Playing bored games is our favorite pastime.
    d. No errors.

33. Choose the sentence that contains an error in usage. If there are no errors, select Choice *D*.
    a. My mom likes to eat hot peppers straight from the jar.
    b. Georgia taught herself to cook by watching her grandmother in the kitchen.
    c. Because she couldn't sleep well before her exam, Marisa was particularly exhausted.
    d. No errors.

34. Choose the sentence that contains an error in usage. If there are no errors, select Choice *D*.
    a. My sister loves babysitting infants.
    b. I want to go skiing, but I can't afford it.
    c. Thomas has two knew puppies.
    d. No errors.

35. Choose the sentence that contains an error in usage. If there are no errors, select Choice *D*.
    a. Their are too many kids on the bus.
    b. We ordered four bagels: poppyseed, plain, egg, and onion.
    c. They're going to call us back soon.
    d. No errors.

36. Choose the sentence that contains an error in usage. If there are no errors, select Choice *D*.
    a. Whose coming to your birthday party?
    b. The effects of the medication wear off after three hours.
    c. When I get coffee, I put a lot of milk in it.
    d. No errors.

37. Choose the sentence that contains an error in usage. If there are no errors, select Choice *D*.
   a. I hate when theres a long line at the bank.
   b. My favorite holiday is Thanksgiving, but I also love Christmas.
   c. I have to get a gift for Aunt Linda.
   d. No errors.

Spelling

38. Choose the sentence that contains an error in spelling. If there are no errors, select Choice *D*.
   a. It is considerably warmer in the dormitory than it is in the lecture hall.
   b. The example illustrates her point well.
   c. What is your prefrence: coffee or tea?
   d. No errors.

39. Choose the sentence that contains an error in spelling. If there are no errors, select Choice *D*.
   a. I heard a rumor that there was a fire in the laboratory.
   b. On Wendesday, I have to go to the recycling center with my coworker.
   c. Paula will rescind her offer if we don't accept it by September.
   d. No errors.

40. Choose the sentence that contains an error in spelling. If there are no errors, select Choice *D*.
   a. Betty is responsible for securing everyone's valuables.
   b. The circumfrence of the circle is 50 centimeters.
   c. It is necessary to use scissors to cut the construction paper.
   d. No errors.

41. Choose the sentence that contains an error in spelling. If there are no errors, select Choice *D*.
   a. The extravagant party was successful.
   b. Pedro's grandmother is starting to become noticeably forgetful.
   c. Gail told her sister that it was ridiculous to assume her story was believable.
   d. No errors.

42. Choose the sentence that contains an error in spelling. If there are no errors, select Choice *D*.
   a. Please schedule the appointment at your earliest convenience.
   b. My preferred route to school avoids the highway.
   c. Kim's appetite was supressed while she was sick.
   d. No errors.

43. Choose the sentence that contains an error in spelling. If there are no errors, select Choice *D*.
   a. After I bought the newspaper, the cashier gave me a quarter and two nickels.
   b. Our principal has a very easygoing temperment.
   c. Paula will rescind her offer if we don't accept it by September.
   d. No errors.

44. Choose the sentence that contains an error in spelling. If there are no errors, select Choice *D*.
   a. Daniel is refusing to acknowledge his responsibility in escalating the argument.
   b. What is the most purposeful equipment for that task?
   c. That policy is referring to amateur participants.
   d. No errors.

45. Choose the sentence that contains an error in spelling. If there are no errors, select Choice *D.*
   a. During which occurrance was the toddler acting the most outrageous?
   b. Approximately how often are you experiencing migraines?
   c. I'm particularly concerned that this gadget is not guaranteed to work.
   d. No errors.

46. Choose the sentence that contains an error in spelling. If there are no errors, select Choice *D.*
   a. The couple was hoping to strengthen their marraige.
   b. Tate was surprised he achieved a perfect score because he was so fatigued.
   c. The atmosphere was somber and people were clearly distraught.
   d. No errors.

47. Choose the sentence that contains an error in spelling. If there are no errors, select Choice *D.*
   a. When I can't pronounce words that I read aloud, I feel embarrassed.
   b. Gerry did a thorough job attaining evidence.
   c. The bulliten board should be hung in a conspicuous location.
   d. No errors.

Composition

48. In English class we had to _____ which was the main character of the text in the novel we were reading.
   a. revolt
   b. dominate
   c. identify
   d. oppose

49. The _____ source means a source that was created at the same time that is being studied or analyzed. Some examples include a diary, recording, or artifact.
   a. accurate
   b. consistent
   c. dissatisfied
   d. primary

50. The comedian was _____. We laughed throughout the entire show.
   a. frequent
   b. hilarious
   c. harsh
   d. scarce

51. When the new dog from next door appeared in Harry's front yard, Harry _____ stuck his hand out to pet him.
   a. portably
   b. cautiously
   c. morally
   d. seldomly

52. Abigail never showed up at the correct time and was therefore not very _____.
   a. reliable
   b. frigid
   c. visual
   d. absurd

53. We couldn't tell the exact number of guests, but we knew it was _____ eighty.
   a. exactly
   b. sadly
   c. approximately
   d. occasionally

54. We loaded the _____ into the ship to be delivered in one week.
   a. riot
   b. morsel
   c. cargo
   d. voyage

55. While staying at the ski resort, we saw a great cliff of snow break off the mountain and discovered that it was a(n) _____.
   a. beverage
   b. jubilee
   c. catapult
   d. avalanche

56. Our vacation was very _____; we sat under palm trees on the ocean, watched kids playing with coconuts, and soaked under the warm, humid sun.
   a. tropical
   b. cold
   c. urban
   d. minor

57. At the _____, they served food and gave speeches.
   a. hospital
   b. prelude
   c. banquet
   d. hangar

58. The queen's _____ was prosperous for the country and lasted for fifty years.
   a. reign
   b. purchase
   c. harbor
   d. oasis

59. She was a(n) _____ because she believed in hope and favorable outcomes.
   a. investigator
   b. optimist
   c. journalist
   d. nihilist

60. The island was very _____; there were only five houses on it, and one had to get there by boat.
    a. formal
    b. colonial
    c. dormant
    d. remote

Answer Explanations

1. A: The choice with an issue is Choice *A*. When two people are named and both possess the same object, the apostrophe-*s* indicating possession should be placed ONLY after the second name, NOT after both names.

2. C: Choice *C* contains an error. The first possessive is *whose*, NOT "who's," a contraction of "who is." It should be noted that in Choice *B*, the plural possessive noun *friends'* is correctly punctuated with an apostrophe following the plural -*s* ending. *Friend's* would indicate a singular possessive noun, like something belonging to one friend. *Friends* would indicate a plural noun not possessing anything.

3. B: The noun "teachers" is not a proper noun, and therefore it should not be capitalized. Therefore, Choice *B* is the correct answer.

4. A: Choice *A* is the correct answer because the cities of Springfield, Alamo, and Moscow should be capitalized. The other answer choices do not have errors in capitalization or punctuation.

5. D: There are no errors in the sentences.

6. D: There are no errors in the sentences.

7. B: There is a comma splice in the sentence: The boys never met, they were strangers. A comma splice occurs when a comma is used to incorrectly join two independent clauses.

8. A: The rule for writing a quotation with another quotation inside it is to use double quotation marks to enclose the outer quotation, and use single quotation marks to enclose the inner quotation. The correct usage is reversed in Choice *A*.

9. D: There are no errors in the sentences. Choice *A* correctly capitalizes the proper nouns. In Choice *B*, the name of the U.S. *Constitution* is capitalized, the related adjective *constitutional* is NOT capitalized. This is correct. Choice *C* correctly capitalizes the specific historical era.

10. B: Titles of very high-ranking government officials are capitalized even without their name following the official title in cases where a specific person is still being referred to. Therefore, "president" is capitalized in Choice *C*, as it should be, but it should not be capitalized in Choice *B*, as it is now, because the sentence is just referring to the position in general, not a specific president. Governor, in Choice *A*, is not capitalized, which is correct, because this lower-level title is not addressing the person specifically as it is in the following: "Do you want more coffee, Governor?"

11. C: Choice *C* is a question so it should be punctuated with a question mark at the end, not a period. The other sentences do not contain errors.

12. C: Choice C is missing a comma after *friend* and before *Selma*. The other sentences are properly punctuated. A semicolon is used to join two independent clauses, and a colon is used before a list.

13. C: Choice *C* contains a word choice usage error. *Affect* is a verb that means to impact something. The word *effect* can be used as either a noun or a verb. *Effect*, when used as a noun, means the result or outcome of a cause. When used as a verb, it means to cause or bring about. Here, the turbulence impacted the speaker's ability to read so it *affected* the ability to read.

14. A: Choice *A* uses incorrect usage, therefore it is the correct answer. The word "too" should be replaced with the word "two."

15. D: There are no errors in usage in Choices *A*, *B*, or *C*.

16. D: There are no errors in usage in Choices *A*, *B*, or *C*.

17. B: Choice *B* contains a usage error. "Please call Sharon and I after you get home from your trip" should read: "Please call Sharon and me after you get home from your trip."

18. A: Choice *A* contains a usage error because the plural of "moose" is "moose."

19. A: Choice *A* is the correct answer because there is a usage issue in Choice *A*. The word "Its" should be changed to "It's," a contraction for "It is."

20. D: There are no errors in usage in Choices *A*, *B*, or *C*.

21. A: Choice *A* contains an error in usage. "Everyday" should be "every day." *Everyday* is an adjective that is used when referring to something commonplace, ordinary, or typical. The two-word phrase *every day*, as a phrase, means "each day."

22. A: Choice *A* contains a word choice usage error. The word *tale* means a story. The intended word was *tail*, which is the hind body part or extension on an object resembling the animal's body part.

23. B: Choice *B* contains a dangling modifier, so it contains the error. Dangling modifiers modify a phrase or word that is not clearly found in the sentence, which renders them unattached. In the sentence *Having finished the soup, Walter washed the bowl*, "Having finished" is a participle expressing action. However, the doer is not the soup, it is Walter. The sentence should be: Having finished the soup, Walter washed the bowl.

24. D: There are no errors in usage in Choices *A*, *B*, or *C*.

25. A: Choice *A* contains a subject/verb agreement error. "Hopes" is plural, so the verb must be plural as well. The correct usage would be: "Our hopes were dashed after the agent told us the tickets were sold out."

26. B: There is a word choice usage error in Choice *B*. The word "accept" is used incorrectly instead of "except." *Accept* means to agree or consent to something, while *except* means other than.

27. C: There is a usage error in Choice *C*. The word "Their" should be "There." "Their" is used to indicate possession of something owned or attributed to a group of people.

28. D: There are no errors in usage in Choices *A*, *B*, or *C*.

29. C: Choice *C* has an error is usage because it does not adhere to parallel structure. The gerund form of the verb (*riding*) is needed. It should read: She likes hiking, dancing, and riding her bike.

30. A: There is an error in word choice usage in Choice *A*. *Then* should be *than*. *Then* is typically used as an adverb to describe that something has happened next in sequential order or because of a conditional situation. For example: We ate the cake and *then* played party games. Or: If it snows, then we will need to shovel. *Than* is a conjunction used make a comparison. For example: I like ice cream more than cake.

31. D: There are no errors in usage in Choices *A*, *B*, or *C*.

32. C: Choice *C* is the correct answer because there is a usage error in the sentence. The word "bored" should be changed to "board." The correct terminology is "board games," not "bored games."

33. D: There are no usage errors in Choices *A*, *B*, or *C*.

34. C: Choice *C* is correct because it has an issue in usage. The word "knew" should be replaced with "new."

35. A: Choice *A* is correct because there is an error in usage with the word "Their," which should be replaced with "There."

36. A: Choice *A* is correct because there is a usage error. The word "Whose" should be replaced with "Who's," which is a contraction for "Who is."

37. A: Choice *A* is correct because there is an error in usage with the word "theres." This word should be replaced with "there's," which is a contraction for "there is."

38. C: There is a spelling error in Choice *C*. "Prefrence" should be "preference."

39. B: There is a spelling error in Choice *B*. "Wendesday" should be "Wednesday."

40. B: There is a spelling error in Choice *B*. "circumfrence" should be "circumference."

41. D: There are no spelling errors.

42. C: There is a spelling error in Choice *C*. "Supressed" should be "suppressed."

43. B: There is a spelling error in Choice *B*. "Temperment" should be "temperament."

44. D: There are no spelling errors.

45. A: There is a spelling error in Choice *A*. "Occurance" should be "occurence."

46. A: There is a spelling error in Choice *A*. "Marraige" should be "marriage."

47. C: There is a spelling error in Choice *C*. "Bulliten" should be "bulletin."

48. C: In English class we had to *identify* which was the main character of the text. Identify means to recognize or label something.

49. D: A *primary* source is one that was created at the same time that is being studied or analyzed. Primary means first or original.

50. A: The comedian was *hilarious*. The surrounding context shows that the speaker laughed at the comedian's show, which implies that the comedian was *hilarious*, or funny.

51. B: Harry *cautiously* stuck his hand out to pet him. To be cautious of something means to be hesitant or careful. Harry doesn't know if the new dog is friendly or mean, so he slowly, carefully, sticks out his hand.

52. A: Abigail was not very *reliable*. Reliable means trustworthy or dependable, so if Abigail does not show up on time, then she is not very reliable. **53. C:** We knew it was *approximately* eighty. The word approximately means almost exact or almost accurate. Since they couldn't tell the *exact* number of guests, they guessed at the approximate number of guests.

54. C: We loaded the *cargo* into the ship. Cargo is baggage or shipment that a ship carries for delivery.

55. D: We discovered that it was an *avalanche*. An avalanche is a sudden rush of something, especially snow coming from a mountain. A ski resort surrounded by mountains is very likely to witness a snow avalanche.

56. A: The word *tropical* signifies characteristics of a hot, humid, temperate climate usually in the tropics. Tropical is the best word here given the context clues of weather, animals, and activities related to a tropical atmosphere.

57. C: The best word is *banquet*. Looking at the context clues, we have to choose a noun that involves a place where people serve food and give speeches, such as a banquet.

58. A: The queen's *reign* was prosperous for the country and lasted for fifty years. The word reign means the period of time in which a king or queen occupies a throne.

59. B: She was an *optimist* because she believed in hope and favorable outcomes. An optimist believes that everything will turn out in good favor.

60. D: The island was very *remote*. Remote means that something is secluded or far away from civilization.

Greetings!

First, we would like to give a huge "thank you" for choosing us and this study guide for your HSPT exam. We hope that it will lead you to success on this exam and for your years to come.

Our team has tried to make your preparations as thorough as possible by covering all of the topics you should be expected to know. In addition, our writers attempted to create practice questions identical to what you will see on the day of your actual test. We have also included many test-taking strategies to help you learn the material, maintain the knowledge, and take the test with confidence.

We strive for excellence in our products, and if you have any comments or concerns over the quality of something in this study guide, please send us an email so that we may improve.

As you continue forward in life, we would like to remain alongside you with other books and study guides in our library, such as:

PSAT 8/9: amazon.com/dp/1628456825

PSAT: amazon.com/dp/1628457384

SAT APEX: amazon.com/dp/1628458224

We are continually producing and updating study guides in several different subjects. If you are looking for something in particular, all of our products are available on Amazon. You may also send us an email!

Sincerely,
APEX Test Prep
info@apexprep.com

FREE

Free Study Tips DVD

In addition to the tips and content in this guide, we have created a FREE DVD with helpful study tips to further assist your exam preparation. **This FREE Study Tips DVD provides you with top-notch tips to conquer your exam and reach your goals.**

Our simple request in exchange for the strategy-packed DVD is that you email us your feedback about our study guide. We would love to hear what you thought about the guide, and we welcome any and all feedback—positive, negative, or neutral. It is our #1 goal to provide you with top quality products and customer service.

To receive your **FREE Study Tips DVD**, email freedvd@apexprep.com. Please put "FREE DVD" in the subject line and put the following in the email:

    a. The name of the study guide you purchased.

    b. Your rating of the study guide on a scale of 1-5, with 5 being the highest score.

    c. Any thoughts or feedback about your study guide.

    d. Your first and last name and your mailing address, so we know where to send your free DVD!

Thank you!